T0275737

Adaptive Radar Resource Management

Peter W. Moo
Zhen Ding

AMSTERDAM • BOSTON • HEIDELBERG • LONDON
NEW YORK • OXFORD • PARIS • SAN DIEGO
SAN FRANCISCO • SINGAPORE • SYDNEY • TOKYO
Academic Press is an imprint of Elsevier

Academic Press is an imprint of Elsevier
125 London Wall, London, EC2Y 5AS, UK
525 B Street, Suite 1800, San Diego, CA 92101-4495, USA
225 Wyman Street, Waltham, MA 02451, USA
The Boulevard, Langford Lane, Kidlington, Oxford OX5 1GB, UK

Notices
Knowledge and best practice in this field are constantly changing. As new research and experience broaden our understanding, changes in research methods, professional practices, or medical treatment may become necessary.

Practitioners and researchers must always rely on their own experience and knowledge in evaluating and using any information, methods, compounds, or experiments described herein. In using such information or methods they should be mindful of their own safety and the safety of others, including parties for whom they have a professional responsibility.

To the fullest extent of the law, neither the Publisher nor the authors, contributors, or editors, assume any liability for any injury and/or damage to persons or property as a matter of products liability, negligence or otherwise, or from any use or operation of any methods, products, instructions, or ideas contained in the material herein.

Library of Congress Cataloging-in-Publication Data
A catalog record for this book is available from the Library of Congress

British Library Cataloguing in Publication Data
A catalogue record for this book is available from the British Library

For information on all Academic Press publications
visit our website at http://store.elsevier.com/

ISBN: 978-0-12-802902-2

Working together
to grow libraries in
developing countries

www.elsevier.com • www.bookaid.org

DEDICATION

To Michelle, Karolina, and Evan   — PWM

To Kebing, Shangzi, and Josephine   — ZD

CONTENTS

ACKNOWLEDGMENTS

The authors are grateful to David DiFilippo for numerous helpful discussions. They also thank Bill Brinson, Bing Yue, and Joseph Chamberland of C-CORE Ottawa for carrying out simulations in Adapt_MFR, and Dana Rakus of DRDC for preparing several figures.

Introduction

The use of phased array antennas has enhanced the flexibility and effectiveness of radar. In particular, phased array technology allows the radar beam to be controlled and adapted almost instantaneously. This flexibility enables the radar to carry out multiple functions simultaneously, such as surveillance, tracking, and fire control, where each function carries out a number of looks. The execution of multiple functions necessitates the study of radar resource management (RRM), which considers the prioritization and scheduling of radar looks, as well as task parameter selection and optimization. RRM is especially important in overload situations, when the radar does not have sufficient time to schedule all requested looks. In this case, the radar scheduler must decide which looks should be scheduled and which should be delayed or dropped. Additionally, for the looks to be scheduled, a start time for each look must be determined.

1.1 THE RADAR RESOURCE MANAGEMENT PROBLEM

In this section, we describe the RRM problem from four perspectives, including radar resources, radar terminology, radar functions, and the presentation of an RRM model.

1.1.1 Radar Resources

A phased array multifunction radar (MFR) performs many functions previously performed by individual, dedicated radars, such as surveillance, tracking, and fire control. The radar performs these functions by actively controlling its beam position, dwell time, waveform, and energy. Details of

Adaptive Radar Resource Management. http://dx.doi.org/10.1016/B978-0-12-802902-2.00001-6

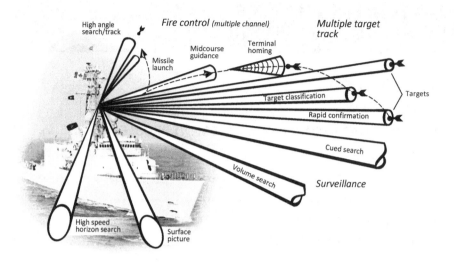

Figure 1.1 Multiple functions of ship-borne radar systems.

general phased array radars can be found in [1–3]. An illustration of the multiple functions is depicted in Figure 1.1.

There are typically several tasks associated with each radar function. All the functions and function tasks are coordinated by the RRM in the radar system. This RRM component is critical to the success of an MFR, since it maximizes the radar resource usage in order to achieve optimal performance, where the optimality is defined according to various cost functions.

There are three major radar resources, as shown in Figure 1.2. The challenge of RRM arises when the radar resources are not sufficient to carry out all function tasks. Lower-priority tasks must encounter degraded performance due to fewer available resources, or the radar may not execute some tasks at all. Each task in the radar requires a certain amount of time, energy, and computational resources. The time is characterized by the tactical requirements, the energy is limited by the transmitter energy, and computational resources are limited by the RRM computer. All of these limitations have impacts on the performance of the radar resource manager.

Among the radar resources, the time budget is the most constraining, since the radar cannot create additional timeline. The energy budget is typically limited by the available power supply and the cooling system. The processing budget is usually the least constraining because of the

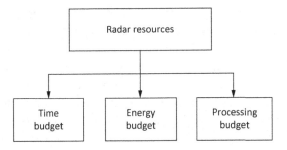

Figure 1.2 Multifunction radar resources.

ever-increasing capability of computer processors. A radar scheduler co-ordinates the usage of all radar resources in carrying out RRM.

1.1.2 Radar Terminology

To state the goal of a radar scheduler accurately, it is necessary to distinguish between a function, a task and a look.

Functions. The radar carries out multiple functions, which include weapon control, target tracking, and surveillance. A detailed description of various functions is given in Section 1.1.3.

Tasks. Each function consists of one or more tasks. For the weapon control function, a task involves the control of an individual weapon. Similarly, for the target tracking function, a task involves the tracking of an individual target. The surveillance function monitors a specified region of interest. A surveillance task may include the monitoring of a subregion within the specified region of interest. The surveillance function can also be thought of as consisting of a single task, where the task involves monitoring the entire region of interest.

Looks. Each task consists of several looks, where a look requires one continuous time interval of finite duration to be completed. For a tracking task, a look is an attempt to update a track by steering the radar in the direction of the expected location of the target. In this case, a look could consist of one or more beam positions of the radar. For a surveillance task, a look could consist of a single beam position or multiple beam positions. Since a look has been defined as requiring a continuous time interval to be completed, it is beneficial to define surveillance looks as being as short in duration as possible. This allows the scheduler the flexibility to interleave looks from multiple tasks.

Each task sends look requests to the radar scheduler. For a target tracking task, a look request may consist of an attempt to update a track at a specified time. The specified time will depend on the time of the track update, the estimated target dynamics, and the tracking model. For all tasks, look requests are sent to the radar scheduler independently. That is, each task makes look requests based only on its own requirements. The role of the radar scheduler is to receive all look requests and formulate a schedule for the radar, under the constraint that at any given time, the radar only executes one look. The radar scheduler must decide whether or not to schedule the look request. For example, if two look requests due to start at the same time are received, the scheduler must decide whether to alter the start times of one or both looks, or to not schedule one of the looks.

1.1.3 Radar Functions

The following functions are carried out by a ship-borne MFR [4].

Horizon search. The objective of horizon search is to detect low-flying targets as soon as they cross the radar horizon. Because these threats are perceived to be one of the major threats to maritime surface ships, horizon search is one of the main functions of the MFR.

Cued search. Other sensors in the ship's sensor suite may detect targets that are not yet being tracked by the MFR. This event may occur due to adverse propagation or other conditions for the MFR, a temporary overload of the MFR time budget, or because the target is outside of the coverage area of the MFR. If the target has not yet been tracked due to adverse propagation or other conditions, the length of the cued search dwell is increased compared to the normal search dwell to improve the probability of detection. The cued search pattern, which depends on the source of the cue, is executed only once. The delay between the detection of the target by the other sensor and the actual transmission in the MFR must be short to keep the search volume, and therefore the load on the time budget, as small as possible.

Confirmation. After a target has been detected in a search dwell and the target is not yet in track, a dwell is transmitted in the direction measured by the search dwell to confirm the presence of a target. A successful confirmation results in the initiation of a track. The delay in the transmission of a confirmation dwell must be short to ensure that the target is still within half of a beamwidth of the direction measured by the search dwell.

Air target track. After tracks have been initiated, air targets are tracked with dedicated dwells. The update rate and the dwell time are adapted to the behavior of the target in such a way that the track is maintained with a minimum load on the time-energy budget.

Weapon track. Targets that have been selected for an engagement are tracked with an update rate that is high enough to guarantee a track accuracy that is required for missile guidance.

Surface-to-air missile (SAM) acquisition. A search pattern is executed to acquire the SAM shortly after launch by the ownship. Knowledge about the SAM trajectory is used to define a pattern that has a high probability of acquisition and requires only limited radar time and energy. After a successful acquisition, a track is initiated.

SAM track. SAMs are tracked to collect information that is required for midcourse guidance and to avoid unnecessary usage of resources due to confirmation or cued search dwells in the direction of the SAM.

Midcourse guidance. SAMs that have been launched against a target require midcourse guidance to the predicted intercept point. The actual information in a midcourse guidance message is dependent on the missile type.

Terminal illumination. In the terminal phase of an engagement using semi-active SAMs, targets must be illuminated by the MFR to enable the missile seeker to lock on the target. It is assumed here that the seeker of the semi-active missile requires illumination dwells at very regular intervals. This requirement results from the synchronous operation of the processor in the missile seeker.

Kill assessment. Shortly before and after the predicted intercept, the update rates of the weapon track and SAM tracks are increased to establish the result of the engagement with a high degree of confidence. The result of this assessment is submitted to the combat system to support the decision of launching other missiles.

Each function has a specific demand for a share of the time budget that is determined by the duration of the function, the average number of dwells per second (the update rate), and the dwell time. For search function, the update rate is determined by the number of beams that are required to scan the search volume (frame) and the time between successive dwells in the same direction (frame time).

For the horizon search function, relatively large deviations from the desired time between successive dwells are allowed to enable more important functions, such as terminal illumination, to be scheduled. As a result, the confirmed detection range will decrease, but this is acceptable during overload conditions.

For the midcourse guidance and tracking functions, the time between the desired and actual transmission time of a dwell is more constrained to avoid a significant degradation of the kill probability of the SAMs and tracking performance, respectively. Finally, as has already been noted, the dwells for terminal illumination must be transmitted synchronously.

1.1.4 Radar Resource Management Model

Due to the nature of RRM for MFRs, a model for RRM will necessarily be complex. A general RRM system model is shown in Figure 1.3. It performs the following steps.

- Obtain a radar mission profile or function setup.
- Generate radar tasks.
- Assign priorities to tasks by using a prioritization algorithm.
- Manage available resources via a scheduling algorithm so that the system can meet the requirements of all radar functions.
- For nonsurveillance tasks, schedule a re-look if a target was not detected. This will depend on the task priority and elapsed time since the last scheduling of the same task.

The radar scheduler takes into account a number of factors, including radar beams, dwell time, carrier frequency, pulse repetition frequency (PRF), and transmitted energy. As can be seen from the above steps, the RRM problem has two key elements: task prioritization and task scheduling. RRM algorithms can address these two elements separately or simultaneously. Task prioritization is an important factor in the task scheduler. The other factor is the required scheduling time, which is decided by the environment, the target scenario, and the performance requirements of radar functions. The required scheduling time may be improved by using advanced algorithms, such as waveform-aided algorithms and adaptive update rate algorithms.

Note that the general sensor management problem is to optimally coordinate the usage of multiple sensors, which is not covered in this book.

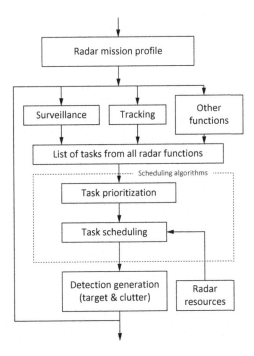

Figure 1.3 A radar resource management model.

1.2 OUTLINE OF THIS BOOK

The book is divided into six chapters. It is written for both practitioners and researchers, focusing on applied algorithms for real-world challenges. Simulation results are presented throughout to quantify the performance of various scheduling algorithms.

Chapter 2 presents a comprehensive overview of RRM techniques. In Chapter 3, a number of adaptive RRM techniques are compared to a nonadaptive baseline resource manager. Chapter 4 presents techniques to optimize the time-sensitive scheduling of tracking tasks. In Chapter 5, RRM techniques for networked radars are proposed and evaluated. Finally, Chapter 6 presents conclusions and future work.

Overview of RRM Techniques

Adaptive Radar Resource Management. http://dx.doi.org/10.1016/B978-0-12-802902-2.00002-8

2.1 INTRODUCTION

The radar resource management (RRM) algorithms surveyed in this chapter are divided into five categories, with one section devoted to each category. The first three categories are adaptive scheduling algorithms, and the remaining two categories are resource-aided algorithms. When a paper falls into more than one category, it is placed into the most suitable category. Categories 4 and 5 are relevant since a better algorithm is able to achieve the same performance with fewer resources or to achieve a better performance with the same radar resources. Comments are provided for the RRM algorithms in each category.

The five categories are:

1. artificial intelligence (AI) algorithms (Section 2.2);
2. dynamic programming (DP) algorithms (Section 2.3);
3. Q-RAM algorithms (Section 2.4);
4. waveform-aided algorithms (Section 2.5); and
5. adaptive update rate algorithms (Section 2.6).

In Section 2.7, the Naval Research Laboratory (NRL) benchmark problems are defined and solutions proposed to date are reviewed. Finally, a summary is presented in Section 2.8.

2.2 ARTIFICIAL INTELLIGENCE ALGORITHMS

In this category, 15 papers are noted [7–21]. The papers cover neural network approaches [7–9], expert system approaches [10, 11], and fuzzy logic approaches [12–18]. An entropy approach for radar scheduling is also discussed [21]. Paper [22] belongs to both the AI category and the waveform-aided algorithm category. It will be discussed in the waveform algorithm category in Section 2.5.

2.2.1 Neural Networks
Neural networks (NNs) are used for key elements of RRM: using classification NNs for task prioritization and optimizing NNs for task scheduling.

2.2.1.1 Task Prioritization
Classification NN algorithms are primarily used for assignment of priorities to tasks. The input is all required radar tasks, and the constraints are radar time and energy budgets. Optimization could be the minimization of radar resources, given the search, track and engagement performance requirements, or the maximization of the performance by using the available radar resources.

Komorniczak [7, 8] proposed an NN priority assignment algorithm. In this algorithm, a feature vector was the input to multi-layer neurons. A training data set was used to adjust weights of the NN. In the application phase, the trained NN generated the priorities based on all given targets feature data. The arbitrary nonlinear mapping capability of the NNs was utilized.

The mapping provides target prioritization values, which classify radar targets into different levels. This is necessary when a lot of targets are competing for radar resources. Accordingly, radar resources are first given to those targets with higher priority. For example, the following target features can be used:

- membership (friend or foe);
- range;
- radial velocity;
- azimuth; and
- acceleration.

Non-numerical features of the targets are transformed to numerical values, which determine the input vector in the target prioritization process. All the features are put into a joint vector as follows:

x_1 membership: friend ($x_1 = 0$), foe ($x_1 = 1$);
x_2 range (km);
x_3 radial velocity of the target (m/s);
x_4 azimuth (°); and
x_5 acceleration of the target (m/s^2).

As shown in Figure 2.1, the components of the x vector are multiplied by weights in the NN block. Then the output u is calculated as a weighted sum, that is:

$$u = \sum_{i=1}^{5} w_i x_i. \tag{2.1}$$

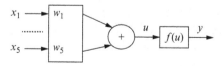

Figure 2.1 The priority assignment module [9].

The track priority output is calculated by a nonlinear activation function $f(u)$:

$$f(u) = \frac{1}{1 + e^{(-bu)}}. \tag{2.2}$$

This function is widely used since it offers continuous priority values in $[0, 1]$. The slope of the function depends on the parameter b. For parameter $b \to \infty$, the function becomes:

$$f(u) = \begin{cases} 1, & u > 0 \\ 0.5, & u = 0 \\ 0, & u < 0 \end{cases}. \tag{2.3}$$

The weight coefficients are generated by a learning method with back propagation. This method relies on minimizing the mean square error. It can be defined as:

$$q = \frac{1}{2} \sum_{j=1}^{N} (\delta^{(j)})^2, \tag{2.4}$$

where

$$\delta^{(j)} = z^{(j)} - y^{(j)}, \tag{2.5}$$

and $z^{(j)}$ is the requested value of the target rank in the jth step of learning. The term $y^{(j)}$ is the output value of the target calculated in the jth step of learning for $w^{(j)}$ weight coefficients, that is:

$$y^{(j)} = f\left(\sum_{i=1}^{5} w^{(j)} x_i^{(j)} \right). \tag{2.6}$$

The parameter N is the number of learning pairs $< x^{(i)}, z^{(i)} >$, and U is a learning set, so that:

$$U = \{< x^{(1)}, z^{(1)} >, \ldots, < x^N, z^N >\}. \tag{2.7}$$

According to the gradient method of minimizing error q, the weights are calculated on the basis of the learning data set:

$$w_i^{(j+1)} - w_i^{(j)} = \Delta w_i^{(j)} = -\eta \frac{\delta q^{(j)}}{\delta w^{(j)}}, \tag{2.8}$$

where η is a learning coefficient.

The weight selection algorithm ensures the minimization of the error q for the established learning set U. Since the learning method is compatible with the nonlinear neuron model and the nonlinear neuron learning algorithm [8], the system has the ability to generalize the target rank and assign the priorities for other targets. The next stage consists of verification of the module in the RRM.

2.2.1.2 Task Scheduling

Optimizing NN algorithms are used for the task scheduling such as pulse scheduling. Izquierdo-Fuente [9] used a Hopfield NN to optimize the radar pulse scheduler and described the general problem, defined the network, and selected the criterion to design the weights. This type of NN does not need a training data set; however, it needs an energy function abstracted from the scheduling problem. This energy function determines the convergence of the network to a solution of proper assignments. A simulation with five targets was carried out to demonstrate the proposed algorithm. However, this approach tends to converge to local as opposed to global minimum solutions. Also, the convergence rate is slow, particular with a large number of targets.

2.2.1.3 Comments

NN algorithms were proposed for both task prioritization and task scheduling. In particular, classification NNs can be useful in RRM and other radar applications such as track classifications. There is no report of NN algorithms being implemented in any prototype or operational radar systems. However, since NNs are very effective in many classification applications, this approach may be useful for target prioritization. One known issue is that generating the learning data sets is not a trivial job, which can significantly impact the effectiveness of the NN algorithms.

2.2.2 Expert System

2.2.2.1 Description of the Expert System Approach

Vannicole [10] and Pietrasinski [11] proposed an expert system with an information database. A high-level expert system diagram is shown in Figure 2.2. This diagram is useful for radar parameter selection, task prioritization, and scheduling. For example, the expert system can be simplified to a scheduler with only a few rules.

The authors presented a knowledge/rule base system, which controlled the parameters and modes for multi-function radars. The expert system performed a situation assessment of the signal/noise environment followed by appropriately prioritized automatic control of the parameters and modes

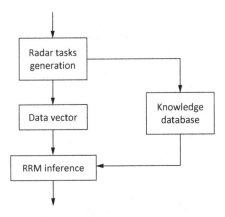

Figure 2.2 Expert system for RRM.

of the radar system. The work involved a radar software development in a simulated environment.

In addition, Vannicole compared a classical solution to a system where an AI approach was used. The two approaches were found to offer similar performance. An application of the expert system was also presented, and its features were described. The proposed method has been tested in an experimental radar system.

2.2.2.2 Comments
Expert systems have not been implemented in real radar systems. Instead, a similar, but more flexible technology of the fuzzy logic has been preferred, which is discussed in the next section.

2.2.3 Fuzzy Logic
2.2.3.1 Fuzzy Logic Approach
References [12–18] describe the use of fuzzy logic to resolve the conflicts of an adaptive scheduler. Here the fuzzy logic allows vague values such as dangerous and friendly to be represented as target priority factors. Fuzzy logic also allows a degree of flexibility to be introduced in tasks for shared resources. Miranda et al. [13, 18] proposed a simulation architecture and decision tree with five fuzzy variables (track quality, hostile, weapon systems, threat, and position). The fuzzy logic approach provided a valid means for prioritizing radar tasks. An adaptive prioritization assignment and fuzzy-reasoning-based algorithm was developed. This algorithm was

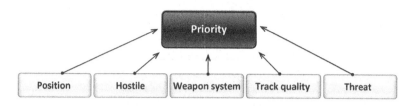

Figure 2.3 Decision tree for target priority assessment.

responsible for ranking tracks and sectors of surveillance in varying tactical environments.

The priorities of targets were evaluated using the decision tree presented in Figure 2.3. The required information to assign a priority was provided by a tracking algorithm. Five different variables provided information for the priority:

1. track quality;
2. hostility;
3. weapon system (describes weapon system capabilities of the platform);
4. threat; and
5. position (of the targets).

Track quality refers to the accuracy of the predicted position of the target with respect to the desired accuracy.

Hostility is a fuzzy variable related to four concepts: range to the targets, absolute target velocity, identity, and the way the target is approaching the radar. Thus, depending on the way the target is approaching the radar platform, its absolute velocity, its range, and its identity, the priority for tracking may vary.

The variable weapon system represents the importance of a target with respect to the weapon systems of the radar. In order to assess its importance, three concepts can be utilized: the identity of the target, the operational range of the weapon systems, and the ratio between the range rate and the absolute velocity of the target.

Threat is the linguistic variable, which represents the degree of threat of a target according to its trajectory and identity. Trajectory combines four fuzzy variables: height, maneuver, absolute velocity, and range rate with respect to the trajectory on which the target is moving. Note that hostile and threat are closely related concepts, but they combine different fuzzy variables.

Table 2.1 Examples of Fuzzy Variables Used in the Assignment of Priorities for Targets [13]	
Fuzzy Variable	**Fuzzy Values**
Track quality	Very low, low, medium-low, medium, medium-high, high, very high
Hostil	Nonhostile, unknown, hostile
Weapon system	Low, medium, high
Threat	Very low, low, medium-low, medium, medium-high, high, very high
Position	Close, medium, far

Finally, position is a linguistic variable whose value is given by the combination of the fuzzy values of the range and azimuth of a target. Fuzzy values are attributed to each variable. Some examples of the fuzzy values are presented in Table 2.1. After evaluation of these variables according to a set of fuzzy rules, the priority of the target is determined.

Stoffel [14] used a dynamic fuzzy logic approach for waveform selection and energy management based on the blackboard architecture. A weapon system simulation test-bed and analytical tool were developed.

In a fuzzy logic processing system, three steps are used: fuzzification, fuzzy rules, and de-fuzzification.

2.2.3.2 Comments
A fuzzy logic controller is implemented in a phased array radar simulator, Adapt_MFR. Experiments showed that this fuzzy logic controller is able to prioritize targets so that they could be scheduled accordingly. The processing speed of the fuzzy controller is fast, making it useful in real radar systems. The fuzzy logic algorithm is used as a baseline against other algorithms.

2.2.4 Entropy
2.2.4.1 Entropy Algorithm
The entropy algorithm was proposed by Berry and Fogg [21]. It used the concept of information entropy for RRM. The approach was particularly appropriate for radar systems dominated by uncertainty and subject to time and resource constraints. The proposed algorithm was applied to the scheduling of track updates in phased array radars.

The objective is to track a number of independent targets using a single multifunction phased array radar by observing them at intermittent times so as to determine their locations and update the tracks. The update rate for each target should ideally be as small as possible so as to maximize the probability of the target being within the beam. On the other hand, the dwell time should be as long as possible in order to maximize the signal-to-noise ratio (SNR), thereby enhancing the probability of detection, while keeping the false alarm (FA) rate low.

However, radar time generally needs to be shared with a number of other targets, as well as the surveillance and weapon guidance tasks. If the update rate or dwell time is too low, then a target may not be detected. Consequently, additional looks will need to be scheduled with higher priorities in order to revisit it. The decision makes trade-offs over time to ensure that the radar's resources are used efficiently, and that as the radar becomes overloaded, its performance degrades.

Suppose there are N targets to be tracked and each target has the following dynamic equation:

$$x(k+1) = F(k)x(k) + v(k). \tag{2.9}$$

The measurement equation is given by:

$$z(k) = H(k)x(k) + w(k), \tag{2.10}$$

where $v(k)$ and $w(k)$ are sequences of zero-mean, white Gaussian noise processes, as normally specified for Kalman filter trackers. Then, $x(k)$ at times $t_k = 0, 1, \ldots$, is a multivariate Gaussian distribution which is estimated by its mean $\hat{x}(k)$ and covariance matrix $P(k)$.

For the purpose of beam scheduling to maintain tracks, the interest is in the elements of the covariance matrix representing the error in target azimuth and elevation, that is, $Q_i(t)$, for the ith target at time t. Then the entropy representing the positional uncertainty is given by:

$$h_i(t) = \frac{1}{2} \log\{4\pi^2 e^2 |Q_i(t)|\}, \tag{2.11}$$

where $|Q_i(t)|$ is the determinant of $Q_i(t)$.

The entropy associated with the joint system of N independent targets at time t is:

$$H(t) = \sum_{i=1}^{N} h_i(t) \tag{2.12}$$

$$= \frac{1}{2} \sum_{i=1}^{N} \log\{4\pi^2 e^2 |Q_i(t)|\}. \tag{2.13}$$

This expression provides a method for quantifying the overall uncertainty associated with the targets, and balancing the resources allocated to them. Alternatively, the optimal control problem could be formulated so as to specify an acceptable level of uncertainty as a constraint. Then the RRM problem becomes one of minimizing resources necessary to maintain that level. This is an appropriate formulation for a set of high-priority targets which must be tracked, with remaining radar resources applied to low-priority targets and other functions.

2.2.4.2 Comments
The entropy algorithm provides an additional approach for track prioritization. In practice, a separate task scheduler is needed. As can be seen, the entropy depends on the filter design for the uncertain covariance matrix. In real applications, the target dynamics are unknown and, therefore, the entropy calculation would be inaccurate. To be more accurate, this requires an adaptive filter in the tracker implementation, which has not been reported. Also, future work is necessary to see if this algorithm performs better than the algorithms previously described such as the NN and the fuzzy logic.

2.3 DYNAMIC PROGRAMMING ALGORITHMS

In the DP algorithm category, 20 papers are noted [23–42]. Unlike the AI-based approaches, the DP algorithms attempt to solve both the task prioritization and task scheduling problems simultaneously.

2.3.1 An Example
The DP approach can be illustrated by a simple three target scheduling problem. Assume that the radar has 5 s of time resource to allocate to three targets for possible track updates. Each target has submitted a number of proposals on how it intends to spend the radar time. Each proposal gives the cost of the scheduling (c) and the total performance gain (r). Table 2.2 gives the proposals generated.

Each target will only be permitted to act on one of its proposals. The goal is to maximize the overall performance gain resulting from the three allocations of the 5 s. It is also assumed that any unused time of the 5 s is lost, just like in a real radar.

Table 2.2 Track Scheduling Options						
Proposal	c1	r1	c2	r2	c3	r3
1	0	0	0	0	0	0
2	1	5	2	8	1	4
3	2	6	3	9	x	x
4	x	x	2	12	2	7

A brute-force way to solve this is to try all possibilities and choose the best total performance gain. In this case, there are $3 \times 4 \times 2 = 24$ ways of allocating the time. Many of these are infeasible. For instance, the three proposals (#3, #2, and #4) for the three targets cost 6 s. Other proposals are feasible, but very poor, such as proposals #1, #1, and #2, which are feasible but performance gain is only 4.

2.3.2 Computational Challenge
There are some serious disadvantages of the brute-force approach:

- For larger problems, the enumeration of all possible solutions may not be computationally feasible.
- Infeasible combinations cannot be detected *a priori*, leading to inefficiency.
- Information about previously investigated combinations is not used to eliminate inferior, or infeasible, combinations.

Note also that this problem cannot be formulated as a linear problem, for the performance gain is not a linear function of the possible proposals. One of the solutions proposed to this optimization problem is the DP algorithm, which computes the optimal radar resource assignment for all tracks. Due to the computational requirement for the high-dimensional cases, seeking more efficient algorithms is an active area of research. It is also noticed that DP algorithms have become more practical with increased computational power.

2.3.3 Some Dynamic Programming Algorithms
Scala and Moran [23] examined the problem of adaptive beam scheduling to minimize target tracking error with phased array radars. It was shown that this could be formulated as a particular type of DP problem, known as the restless bandit problem. Krishnamurthy and Evans [24] derived optimal and sub-optimal beam scheduling algorithms for electronically scanned array tracking systems. The scheduling problem was formulated as a multi-arm bandit problem involving hidden Markov models (HMMs).

Wintenby and Krishnamurthy [25] proposed a more general optimization approach, which led to a two-timescale scheduling solution, and formulated the slow timescale resource allocation as a DP optimization problem. The radar performance was abstracted into performance measures, defined in terms of predicted track accuracy and track continuity. It was done at a slow timescale, and modeled as a discrete time-constrained Markov chain. A Lagrangian relaxation algorithm was used to optimize the radar dynamic measures of performance (MOP).

Wintenby [26] proposed two approaches for scheduling update and search tasks in a phased array radar system. The first approach was based on DP from operations research theory. The other was a temporal reasoning scheme based on temporal logic with a background in AI. In [28], Elshafei et al. presented a new 0-1 integer programming method for the radar pulse interleaving problem, based on Lagrangian relaxation techniques.

Note that two other optimization algorithms have also been proposed. The analysis by Orman [31] was centered on a coupled-task specification of the radar jobs. The coupled-task scheduler is unique in terms of use of idle time within a radar job to interleave other radar jobs, and to achieve improved usage of the radar time. The algorithm proposed by Duron and Proth [35] was based on the concept of time balance and was implemented in the experimental MESAR system. The performance of these two algorithms was found to be similar. Duron and Proth also proposed a strategy to maximize the number of useful tasks performed, considering their priorities.

The performance evaluation of RRM algorithms has been difficult due to the varied nature of the radar configuration, the target, and clutter situation. Dynamic programming algorithms are exponentially intensive and a lot of effort has been dedicated to develop approximate and faster versions, such as [25, 26]. Proth and Duron [37] defined a formal framework for this real-time scheduling problem, and a local search method was introduced to compute efficient schedules for the radar. Based on a V-shape cost function, this algorithm is a good candidate for real-time radar scheduling. A set of lower bounds for the scheduling problem was also described.

2.3.4 Comments

The DP approach, a nonlinear optimization method, has attracted a lot of attention for adaptive radar control. It provides a promising solution to RRM problems. Compared to the target prioritization algorithms, the DP algorithms include the radar configurations and parameter dimensions, and optimize the overall performance of all the tracks. However, this is at the cost of increased complexity, in both the mathematical formulation and the numerical optimization. Published results to date make several theoretical assumptions, such as, specific radar configuration and large selective regions of radar parameters. In practice, radar design is limited within some physical and practical boundaries, such as energy, dwell time and pulse repetition frequency (PRF). The DP algorithms are in the research stage, and the algorithms need to be studied with realistic radar constraints.

2.4 Q-RAM ALGORITHMS

In this category, several papers are noted [43–46]. Similarly to the DP algorithms, the Q-RAM algorithms solve task prioritization and task scheduling simultaneously.

2.4.1 Introduction

The Q-RAM algorithms are based on the concept of quality of service (QoS). The radar system is optimized to maintain an acceptable level of QoS, which is a cost function of performance. Due to the varied nature of the environment, QoS-based resource management has to be adaptive to environments, such as temperature, noise, etc. Consequently, a whole range of resource constraints, such as power, energy, etc., come into play.

2.4.2 Mathematical Formulation

The basic problem solved by Q-RAM is as follows. Given a set of tasks, assign a setpoint such that the system utility is maximized and no resource utilization exceeds its maximum. Formally, it is written as:

$$\text{Maximize: } \sum_{i=1}^{n} u(v_i), \tag{2.14}$$

subject to:

$$\forall 1 \leq k \leq n, \quad 1 \leq i \leq n, \quad r_{ik} = g_{ik}(v_i), \tag{2.15}$$

$$\forall 1 \leq k \leq n, \quad \sum_{i=1}^{n} r_{ik} \leq r_k^{\max}, \tag{2.16}$$

where $g_{ik}(v_i)$ and $u(v_i)$ are the amount of resource k required and the utility derived for task T_i at a setpoint v_i, respectively. While finding the optimal resource allocation is NP hard, the Q-RAM algorithm uses a concave majorant operation to reduce the number of setpoints that must be considered to find a near-optimal solution.

2.4.3 Some Q-RAM Algorithms
2.4.3.1 A Framework of Q-RAM

A Q-RAM radar management framework (Figure 2.4) has three main blocks [47]:

1. The Q-RAM block is a resource allocation tool that employs fast convex optimization using a combination of heuristics and nonlinear programming. It assigns parameters to the radar tasks after considering a variety of factors, including task importance and the current resource utilization level. Q-RAM minimizes the global system error. This objective can also be viewed as utility maximization.
2. The schedulability envelope block is a precomputed schedulability region. It provides Q-RAM with an analytical model of the scheduling operation. Since Q-RAM is a convex optimization engine, the schedulability envelope is transformed into a convex constraint. Satisfaction of this constraint implies that the task set is schedulable with high probability.
3. The last block is a low-level template-based scheduler that generates the dwell schedule based on the parameters computed by Q-RAM. Because the schedulability envelope is computed offline and without knowledge of the runtime system state, it is only an approximate schedulability test. The template-based scheduler provides feedback to Q-RAM when it is unable to schedule a task, and Q-RAM uses this information to update its scheduling constraint. Similarly, when the scheduler generates a schedule

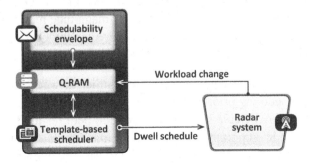

Figure 2.4 Q-RAM framework [47].

that under-utilizes the antenna, it signals Q-RAM to adjust the scheduling constraint.

2.4.3.2 Some Q-RAM Algorithms Based on the Resource Management Framework

Studies have focused on performing feasibility analysis of radar tasks for their given execution times in phased-array radar systems. For example, Kuo et al. [43] proposed a reservation-based approach for real-time radar scheduling. This approach allows the system to guarantee the performance requirement when the scheduling condition holds.

Shih et al. used a template-based scheduling algorithm in which a set of templates was constructed offline, and tasks were fit into the templates at run-time [44, 45]. The templates considered both the timing and power constraints. They also considered interleaving of dwells that allowed beam transmissions (or receptions) on one target to be interleaved with beam transmissions and receptions on another. The space requirements of templates limited the number of templates that could be used, and service classes designed offline determine how QoS operating points were assigned to discrete sets of task configurations across an expected operating range. Goddard et al. [46] addressed real-time back-end scheduling of radar tracking algorithms using a data flow model.

The radar QoS optimization algorithm was based on the work of Q-RAM by Rajkumar et al. [5, 6]. The algorithm used an adaptive QoS middleware framework for QoS-based resource allocation and schedulability analysis in radar systems [48].

In [48], Ghosh et al. proposed an integrated framework for utility maximization and dwell scheduling. Novel concepts such as the scheduling envelope and temporal distance-constrained task model were proposed. Heuristics were used to achieve a two order-of-magnitude reduction in optimization time over the basic Q-RAM approach, allowing QoS optimization and scheduling of a 100-task radar problem to be performed in 700 ms.

A recurring theme in scheduling is the conflict between semantic importance and scheduling priority. Scheduling based on semantic importance alone leads to unpredictable system behavior and poor resource utilization. Here, the semantic importance is defined by the targets threat level. On the other hand, real-time scheduling using earliest deadline first (EDF) or rate monotonic (RM) priorities ignores semantic importance but provides high utilization. The proposed framework reconciled these differences by assigning weights to the tasks based on semantic importance. These weights acted as scaling factors for tracking errors. Since Q-RAM minimizes overall error while ensuring that the system satisfies the scheduling constraint, the system performance would be predictable, and the utilization high while honoring the semantic importance of tasks. More details of the algorithm can be found in [47, 49].

2.4.3.3 Other Q-RAM Algorithms
Gopalakrishnan et al. [50, 51] presented a QoS optimization and dwell scheduling scheme for radar tracking applications. The QoS optimization was performed using the Q-RAM approach. A finite-horizon scheduling algorithm was also proposed. A simulation model of a QoS resource management diagram was proposed and implemented.

Harada et al. [52] proposed a novel control method for fair resource allocation and maximization of the QoS levels of individual tasks. In the proposed adaptive QoS controller, the resource utilization was assigned to each task through an online search for the fair QoS level based on the errors between the current QoS levels and their average. The proposed controller eliminated the need for precise detection of the consumption functions as in conventional feedback control methods. The computational complexity of the proposed method was very low compared to straightforward methods solving a nonlinear problem. The algorithm aimed to maximize system utilities for a soft real-time task set. It is unknown how the algorithm will behave for radar applications.

2.4.3.4 Comments
The optimization goal of Q-RAM is to select a point, or setpoint, in the operation space for each task so as to maximize the global system utility. The utility obtained from a particular setpoint is a function of the setpoint, the environment, and the user defined utility functions. Q-RAM is capable of quickly finding a near-optimal solution. In particular, Q-RAM is designed to work well when the utility curves are concave. This generally holds when there is a law of diminishing returns, where more and more resources are needed to obtain subsequent increases in utility.

The Q-RAM class of algorithms, a nonlinear optimization method, was originally developed in the context of wireless applications, where QoS is a typical performance measure. In radar applications, these algorithms are in the research stage. Currently, the published results present only complicated methods for ideal parameters in a high-dimensional space, leading to an extremely difficult combinatorial problem. For example, an application that has ten QoS dimensions with ten quality levels means that the radar can be configured in numerous ways. It is a research topic that Q-RAM algorithms be evaluated and compared with other algorithms discussed in this chapter.

2.5 WAVEFORM-AIDED ALGORITHMS

2.5.1 Introduction
This class of algorithms assumes that there is a task prioritization and scheduling module. It is focused on improving radar resource requirements by reducing time, energy, and processing budgets via waveform selection. Waveform diversity has been a notable way to optimize radar performance in complex littoral environments with jamming resources. In a multifunction radar (MFR), different waveforms are scheduled for surveillance, detection, tracking, or classification. Waveform selection may use NNs or other optimization techniques. Waveform selection can be a single step or multiple steps ahead. Both fixed and variable waveform libraries have been reported in the literature. Seventeen papers are noted in this category [53–69].

2.5.2 A Neural Network Algorithm
The radar performance factors such as eclipsing, blind velocity, clutter, propagation, and jamming were analyzed by Huizing [22]. These factors were used as the input to a nonlinear function of performance. A multilayer NN was used to model the nonlinear function. A training data set containing pairs of waveform parameters and detection performance was generated with a radar test-bed CARPET. After the training stage, the back propagation network was used to calculate radar detection performance at the selected points in the multidimensional waveform parameter space.

2.5.3 The Waveform Selective PDA Algorithm
Traditional detection and tracking algorithms can be extended to be waveform selective. An example is the waveform selective probabilistic data association algorithm (WSPDA), which is an extension of conventional probabilistic data association (PDA) tracking algorithm [53]. The WSPDA considers a single target in clutter, based on a previous single target (in clutter) Kalman filter tracker [54]. The assumption of an optimal receiver allows the inclusion of transmitted waveform parameters in the tracking subsystem, leading to a waveform selection scheme where the next transmitted

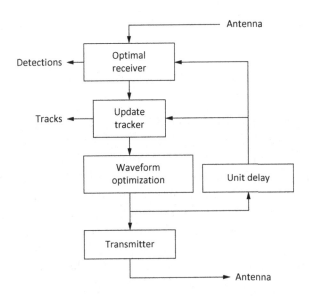

Figure 2.5 Waveform selective PDA tracking system [53].

waveform parameters are selected so as to minimize the average total mean square tracking error at the time step. Semi-closed-form solutions are given to the local one-step-ahead adaptive waveform selection problem for the case of one-dimensional target motion.

The difference between a conventional active transmission tracking system and the new system is the inclusion of a waveform optimization block after the conventional tracking block, as illustrated in Figure 2.5. Thus, the tracking system has active control over the transmitted waveform.

2.5.4 Other Waveform-Aided Algorithms

A waveform-aided interacting multiple model (IMM) tracker was proposed by Howard [55]. The tracker selects a waveform to decrease the dynamic model uncertainty for the target of interest, based on maximization of the expected information obtained about the dynamical model of the target from the next measurement. A design of waveform libraries for target tracking applications was also discussed. Measures of utility of waveforms were defined, and the benefit of adding a particular set of waveforms to the library could be determined.

Suvoroval developed a beam and waveform-scheduling tracker called the Paranoid Tracker [56]. The paper reported an initial study of practical methods for achieving a unification of surveillance and tracking in terms of RRM. The proposed method involved the introduction of permanently existing virtual targets, judiciously placed in the field of view of the radar. The trackers belief in the existence of the virtual or fictitious targets led to the name Paranoid Tracker.

Other waveform-aided detection, tracking, and classification algorithms can be found in references [57–62]. Scala et al. [57] proposed an adaptive waveform scheduling approach for detecting new targets in the context of finite horizon stochastic DP. The algorithm was able to minimize the time taken to detect new targets, while minimizing the use of radar resources. An algorithm to minimize the tracking errors was proposed by Scala et al. [58]. Sowelam and Tewfik [59] suggested that radar waveforms be designed to discriminate between targets. The waveforms minimized decision time by maximizing the discrimination information in the echo signal.

2.5.5 A Literature Survey of Adaptive Radar
In [68], Haykin et al. presented a review of the literature on adaptive radar. The discussion was divided into three groups:

1. controllable parameters for adaptive radar;
2. physical aspects of radar transmission; and
3. detection, tracking, and classification.

2.5.6 A DARPA Research Program on Adaptive Waveform Design for Naval Applications
In 2005, the Defense Advanced Research Projects Agency (DARPA), through NRL, sponsored a research program on adaptive waveform design tailored to naval applications [69]. This project investigates Adaptive Waveform Design for Detecting Low-Grazing-Angle and Small-RCS Targets in Complex Maritime Environments. Team members of this project include: University of Illinois at Chicago, Washington University in St. Louis, Arizona State University, University of Maryland, University of Melbourne, Princeton University, Purdue University, Defence Science and Technology Organization (DSTO), and Raytheon Missile Systems.

The project seeks to achieve substantial improvements in detecting, resolving, and tracking low-grazing-angle (LGA) as well as low-radar cross-section (RCS) targets in maritime and littoral environments under conditions of severe clutter. Transmit waveforms matched to the environment will be integrated with the development of waveform parameters, libraries, realistic models of complex environments, and signal processing methods for optimal waveform selection in real-time to achieve the substantial performance improvements.

2.5.7 Comments

Waveform diversity has been widely studied in the radar and communication community. It provides additional flexibilities for solutions to the RRM problem. Different waveforms have been used in many operational systems. These waveforms are limited within a fixed waveform library. The future trend will be adaptive generation of waveforms on-the-fly, which should be sensitive to the environments and target motions. In addition, finding effective waveforms for specific missions or targets is a challenge. A better waveform saves either time or energy, or both, while maintaining the same level of radar performance.

2.6 ADAPTIVE UPDATE RATE ALGORITHMS

2.6.1 Introduction

An adaptive update rate algorithm is an extension of traditional trackers with uniform update rates. The update rate is closely related to the clutter characteristics, target maneuvering level, and the required tracking performance. Similar to waveform-aided algorithms, the adaptive update rate algorithms optimize the Kalman filter update intervals. Large intervals result in less usage of the radar resources. Therefore, the adaptive update rate algorithms are resource-aided algorithms. Twenty-three papers are collected in this category [70–92].

2.6.2 A Foundation for Adaptive Update Rate Tracking

Daum and Fitzgerald [70] investigated the use of covariance coordinates of various kinds for decoupling Kalman trackers, which resulted in three benefits:

1. reduced computational cost;
2. alleviation of ill-conditioning; and
3. mitigation of nonlinear effects.

This paper provides a foundation for phased array radar tracking where the update rate is variable, and the ill-conditioning and nonlinear issues are more serious.

2.6.3 Adaptive Update Rate IMM-MHT Algorithm

An adaptive update rate tracking algorithm was proposed by Keuk and Blackman [71]. Based on a simple model of a phased array radar, beam scheduling, positioning and radar parameters like SNR, detection threshold were optimized with respect to the computational load. Minimum energy for track maintenance during surveillance was derived.

The revisit time depends on the estimated lack of information regarding the target. Let $\hat{x}(k+1|k)$ be the predicted target state at time $k+1$, along with its covariance $P(k+1|k)$, based on all associated measurements up to time k. Let G denote the major axis of the ellipsoid in u, v space defined by the covariance. Reflecting target maneuverability and position noise, G is an increasing function of the extrapolation time and quite naturally describes the lack of information of the target under track. The relative track accuracy in u, v space is used to calculate the next revisit time $k+1$ from the equation.

$$G(k+1|k) = P_0 B. \tag{2.17}$$

Therefore, the maximum allowed inaccuracy of the track, related to the beamwidth B, is just P_0, the dimensionless track sharpness parameter. Figure 2.6 illustrates the behavior of G as a function of time using (2.17). The sawtooth structure reflects the increase in inaccuracy after a processed report. When the inaccuracy G reaches the threshold, a track update is recommended. After the received observation is processed, G is decreased. Due to initial track uncertainty, we first observe a higher data rate (short revisit times), which afterwards settles when more accurate estimates are available.

In [72], by using adaptive dynamics models and target amplitude information, Koch considered algorithms for efficient control of target revisit intervals, radar beam positions, and energy per dwell with respect to the total sensor allocations and radar energy spent for track maintenance. The performance was evaluated with simulations.

2.6.4 Other Adaptive Update Rate Algorithms

Shin [73] proposed an adaptive update rate IMM algorithm for phased array radar. The purpose of this IMM algorithm was twofold:

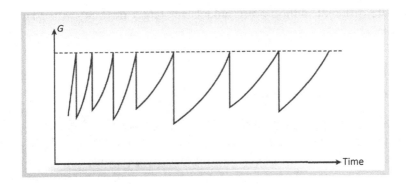

Figure 2.6 Time behavior of track errors.

1. estimate and predict the target states and
2. estimate the level of the dynamic process noise.

The update interval is determined so that the number of track updates per unit time with respect to beam-positioning losses can be reduced. Leung [74] proposed an estimator by solving a formulated dynamic optimization problem using a Hopfield NN. However, the Hopfield NN is not practical due to its low probability of finding the global minimum solution.

Sun-Mog and Young-Hum [75] considered optimal scheduling of track updates to minimize the radar energy, a nonlinear optimal control problem. Keuk and Blackman [71] also proposed a simple model for a multi-target surveillance task. The beam scheduling and radar parameters have been optimized with respect to the radar/computer load. A more realistic rate-based approach was proposed by Tei-Wei [76], where real-time dwell scheduling was considered and significant performance improvement was achieved. This category of algorithms has some overlap with the NRL benchmark solutions discussed in the following section, as they both consider radar parameter optimization for better tracking performance.

2.6.5 Comments
Adaptive update rate algorithms have been used in many radar systems. However, optimal adaptive rate is an ongoing topic. One issue with the adaptive rate is that the tracking parameters are more difficult to optimize than algorithms with uniform or near uniform update rate. This is because the noise matrix in a tracking filter has entities with polynomial functions of the update rate. The noise matrix does not match the real target dynamics when the update rate changes dramatically. Therefore, it is necessary to

formulate different motion noise models for target tracking with adaptive update rate.

2.7 THE NRL BENCHMARK PROBLEMS AND SOLUTIONS

In this chapter, the phased array radar tracking benchmark problems are reviewed. Eighteen papers are discussed in this context [93–110]. The authors of these papers proposed different solutions to the benchmark problems. Remarks are provided for both the benchmark problems and solutions.

2.7.1 The NRL Benchmark Problems

Three benchmark problems were developed by Blair, Watson, and Hoffman and McCabe [93, 99, 109]. The benchmark codes were written in MATLAB and the testing tracking algorithms were to be coded in MATLAB, strictly complying with the input/output format.

The first benchmark problem involves beam-pointing control of phased array radar against highly maneuvering targets. This benchmark includes the effects of target amplitude fluctuations, beam-shape, missed detections, finite resolution, target maneuvers, and track loss.

The second benchmark problem is an extension of the first in that it considers the presence of electronic counter-measurement (ECM) and FAs.

The third benchmark considers closely spaced objects (CSO) and sea-induced multi-path. It also includes the simulation of two additional sensors: the infrared search and track (IRST) and the precision electronic support measurement (PESM).

The benchmark package has the following features:

- funded by NRL in 1994 (Benchmark 1) [93], 1995 (Benchmark 2) [99], and 1999 (Benchmark 3) [110];
- 60 × 60 array (3600 elements, Benchmarks 1, 2, and 3);
- 4 GHz mono-pulse radar (Benchmarks 1, 2, and 3);
- 6 maneuvering targets (Benchmarks 1 and 2) and 12 maneuvering targets (Benchmark 3);
- MOP is the weighted average of energy and time (Benchmarks 1, 2, and 3);

- consider FAs (Benchmarks 2 and 3);
- consider ECM, standoff jammer (SOJ), and range gate pull off (RGPO) (Benchmarks 2 and 3);
- consider sea-surface-induced multi-path and CSO (Benchmark 3); and
- simulation of other sensors such as IRST and PESM (Benchmark 3).

2.7.2 Solutions to the Benchmark Problems

The flow charts of the benchmark problems are depicted in Figures 2.7 and 2.8. Each benchmark participant provides a tracking algorithm denoted Tracking Algorithm. For each experiment, the tracking errors, radar energy, and time are saved. After the last experiment of the Monte Carlo simulation, the average tracking errors, average radar energy per second, and average radar time per second are computed for maintained tracks and the percentage of lost tracks is also computed. A track is considered lost if the distance between the true target position and the target position estimate exceeds 1

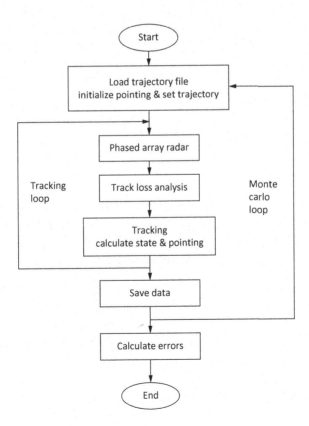

Figure 2.7 Flow chart for Benchmarks 1 and 2 [93].

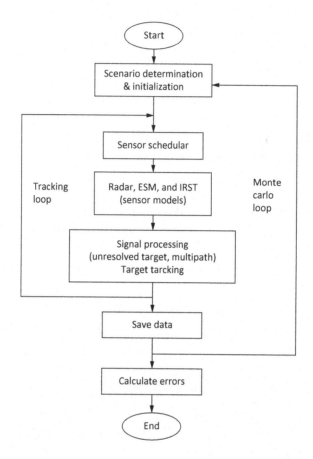

Figure 2.8 Flow chart for Benchmark 3 [109].

beamwidth in angle or 1.5 range gates. A constraint of 4% is to be imposed on the number of lost tracks. When FAs and ECM are present in an actual radar system, algorithms for reacquiring the target and coasting the target tracks through jamming signals are required in order to maintain a track.

Since the benchmark problems were proposed, many solutions, such as the IMM-Multiple Hypothesis Tracker (MHT), alpha-beta tracker, the IMM estimator with probabilistic data association filter (IMMPDAF), adaptive IMM and CSO tracker, have been proposed [100–102, 108, 110]. The results of these solutions are summarized below.

2.7.2.1 Solutions to Benchmark 1

More results were published for Benchmark 1 than for the other two benchmark problems. Note that longer update rates correspond with better solutions. Alpha-beta achieved an update rate of 0.85 s, and the Kalman filter achieved an update rate of 1 s. The H-infinity filter also resulted in an update rate of about 1 s. The update rates of the IMM algorithms were found to be as follows:

- Two-model IMM: 1.3 s.
- Three-model IMM: 1.5 s.
- Adaptive three-model IMM: 2.3 s.

2.7.2.2 Solutions to Benchmark 2
For Benchmark 2, the adaptive Kalman filter achieved an update rate of 1.2 s, and IMMPDAF achieved an update rate of 2.4 s. IMM-MHT results have been published, but there was no detailed results comparison. Computationally, IMM-MHT takes five times longer than the IMMPDAF.

2.7.2.3 A Solution to Benchmark 3
Benchmark 3 was not publicly available, and there is only one publication to date [110]. In the solution presented in [110], Sinha et al. proposed a few algorithms, which are highlighted below.

- Developed both detection and tracking algorithms with enhanced performance, comparing with conventional detection, and tracking algorithms.
- Proposed a modified version of the maximum likelihood (ML) angle estimator, which could produce two measurements from a single detection, and a modified generalized likelihood ratio test (GLRT) to detect the presence of two unresolved targets.
- Sea-surface-induced multipath can produce a severe bias in the elevation angle measurement when the conventional monopulse ratio angle extractor method is used. A modified version of the ML angle extractor was proposed, which produced nearly unbiased elevation angle measurements and significantly improved the track accuracy. Efficient radar resource allocation algorithms for two closely spaced targets and targets flying close to the sea surface were also proposed.
- Finally, an IMMPDAF was used. It was found that a two-model IMM-PDAF performs better than the three-model version used in the previous benchmark. Also, the IMMPDAF with a coordinated turn model had better performance than that using a Wiener process acceleration model.

The presented signal processing and tracking algorithms, operating in a feedback manner, provide an effective solution to Benchmark 3.

2.7.3 Comments

In the benchmark problems, the update rate and energy are requested on the basis of need. There is no consideration given as to whether the required resources are available. This is a drawback for all the benchmark problems.

Benchmark 3 is the most practical among the benchmark problems for naval applications. However, all three NRL benchmarks are designed for target tracking evaluation, which only considered tracking tasks without track prioritization. It is therefore only part of the RRM solution. In particular, it does not consider beam scheduling over search and prioritized tracking tasks. Each track from the tracker simply asks for a detection, which consumes some radar resources.

The proposed solution to Benchmark 3 does not use the IRST and the PESM for target detection. Further investigation is essential, including the incorporation of additional sensors to enhance the RRM performance.

All the benchmark problems use a simplified performance criterion. Advanced MOP could be explored.

2.8 SUMMARY

In this chapter, a survey of RRM algorithms is presented. The algorithms are divided into five categories: AI algorithms, DP algorithms, QoS resource allocation management (Q-RAM) algorithms, waveform-aided algorithms, and adaptive update rate algorithms. Among the five categories, the first three categories are adaptive scheduling algorithms and the remaining two categories are resource-aided algorithms. Also discussed are the NRL phased array radar benchmark problems and solutions.

CHAPTER 3

Comparison of Adaptive and Nonadaptive Techniques

This chapter compares the performance of an adaptive radar resource management (RRM) technique to that of a nonadaptive technique, using modeling and simulation. Section 3.1 specifies the metrics that are computed to quantify performance. Section 3.2 describes the simulation tool Adapt_MFR that is used in the comparison. In Section 3.3, the adaptive RRM technique is described, and the details of its adaptive prioritization, scheduling and track update intervals are quantified. Section 3.4 presents the scenario under consideration and the results of the comparison.

3.1 PERFORMANCE METRICS

RRM involves many components of the radar. Its performance is quantified by overall radar performance. To be specific, RRM is evaluated with respect to the scheduler, the detector, and the tracker, where detection and tracking are the two primary multifunction radar (MFR) functions.

3.1.1 Scheduler Performance Metrics

Scheduler performance metrics are those which are directly related to how timely the multifunction beams are scheduled. These metrics are as follows.

Adaptive Radar Resource Management. http://dx.doi.org/10.1016/B978-0-12-802902-2.00003-X

Maximum delay (MD) is the largest delay of all scheduled beams. The MD could be applied to different functions, such as Surveillance MD (SMD) and Tracking MD (TMD).

Accumulated delay (AD) is the summation of delays of all scheduled beams. The AD could be applied to different functions, such as Surveillance AD (SAD) and Tracking AD (TAD).

Ratio of scheduling (RS) is the ratio of the number of scheduled beams to the total number of beams of the radar mission.

Surveillance occupancy (SO) is defined as the ratio of the surveillance time to the total time.

Tracking occupancy (TO) is defined as the ratio of the tracking time to the total time.

When computing SO and TO, track confirmations are considered as part of the detection process, confirming detections to decide whether a track shall be initialized for the target.

3.1.2 Detection Performance Metrics

Probability of detection (P_d) is defined as P_d of specific targets.

Frame time (FT) is defined as the revisit time of the first detection beam position. Typically the radar comes back to the first detection beam after finishing all detection beams. The FT can be defined for a specific region when there are regions of different priorities.

3.1.3 Tracker Performance Metrics

Target indication accuracies: are a measure of the error between the true target positions and the estimated track positions. Target indication accuracy is measured for range, azimuth, and elevation.

$$\text{TIA}_R(j, i) = \hat{R}(j, i) + \hat{\dot{R}}[t(j, i+1) - t(j, i)] - R(j, i+1), \qquad (3.1)$$

$$\text{TIA}_\theta(j, i) = \hat{\theta}(j, i) + \hat{\dot{\theta}}[t(j, i+1) - t(j, i)] - \theta(j, i+1), \qquad (3.2)$$

$$\text{TIA}_\phi(j, i) = \hat{\phi}(j, i) + \hat{\dot{\phi}}[t(j, i+1) - t(j, i)] - \phi(j, i+1), \qquad (3.3)$$

where

- TIA = target indication accuracy per target measurement (m or rad);
- R = range (m);

- θ = azimuth (rad);
- ϕ = elevation (rad);
- j = target index;
- i = measurement index;
- $\hat{x}$ = x estimate (m or rad); and
- $\dot{x}$ = x rate (m/s or rad/s).

Aggregate target indication accuracies per target are obtained by taking the mean and standard deviation of the target indication accuracies for individual targets. These values show how well individual targets are being tracked.

$$\overline{\mathrm{TIA}_R}(j, i) = \frac{\sum_{i=1}^{I} \mathrm{TIA}_R(j, i)}{I}, \qquad (3.4)$$

$$\overline{\mathrm{TIA}_\theta}(j, i) = \frac{\sum_{i=1}^{I} \mathrm{TIA}_\theta(j, i)}{I}, \qquad (3.5)$$

$$\overline{\mathrm{TIA}_\phi}(j, i) = \frac{\sum_{i=1}^{I} \mathrm{TIA}_\phi(j, i)}{I}, \qquad (3.6)$$

$$\mathrm{TIA}_{\sigma_R}(j) = \sqrt{\frac{\sum_{i=1}^{I} (\overline{\mathrm{TIA}_R}(j, i) - \mathrm{TIA}_R(j, i))^2}{I - 1}}, \qquad (3.7)$$

$$\mathrm{TIA}_{\sigma_\theta}(j) = \sqrt{\frac{\sum_{i=1}^{I} (\overline{\mathrm{TIA}_\theta}(j, i) - \mathrm{TIA}_\theta(j, i))^2}{I - 1}}, \qquad (3.8)$$

$$\mathrm{TIA}_{\sigma_\phi}(j) = \sqrt{\frac{\sum_{i=1}^{I} (\overline{\mathrm{TIA}_\phi}(j, i) - \mathrm{TIA}_\phi(j, i))^2}{I - 1}}, \qquad (3.9)$$

where

- TIA = target indication accuracy per target measurement (m or rad);
- $\overline{\mathrm{TIA}}$ = mean target indication accuracy of all measurements of one target;
- R = range (m);
- θ = azimuth (rad);
- ϕ = elevation (rad);
- j = target index;
- i = measurement index; and
- I = number of measurements.

Aggregate target indication accuracies for all target is obtained by taking the geometric mean of all individual target TIA means. This value shows how well the tracker is performing in general for all targets. The aggregates are measured for range, azimuth, and elevation.

$$GM\_\overline{TIA_R}(j, i) = \sqrt[J]{\prod_{j=1}^{J} \overline{TIA_R}(j)}, \qquad (3.10)$$

$$GM\_\overline{TIA_\theta}(j, i) = \sqrt[J]{\prod_{j=1}^{J} \overline{TIA_\theta}(j)}, \qquad (3.11)$$

$$GM\_\overline{TIA_\phi}(j, i) = \sqrt[J]{\prod_{j=1}^{J} \overline{TIA_\phi}(j)}, \qquad (3.12)$$

where

- $\overline{TIA}$ = mean target indication accuracy of all measurements of one target;
- $GM\_\overline{TIA}$ = geometric mean target indication accuracy for all targets;
- R = range (m);
- θ = azimuth (rad);
- ϕ = elevation (rad);
- j = target index; and
- I = number of measurements.

Track completeness is defined as follows:

$$TC = \frac{\text{Total time for which any confirmed track number is allocated to the target}}{\text{Total time that the target is within defined detection region}}$$

$$(3.13)$$

where the time interval considered in the numerator of the above expression starts at the latest of either:

- the time that an confirmed track for the target is first initiated, or
- the time at which the confirmed track enters the defined target detection region if the target is being tracked before it enters the region,

and ends at the earliest of either:

- the time that a confirmed track with the highest track number for the target is terminated, or
- the time at which the confirmed track leaves the defined target detection region.

Track continuity is the number of track breakups per a chosen time period of evaluation for the same known object.

False track rate is defined to be the average number of false tracks per day, where false tracks are any tracks not associated with a known object.

3.2 ADAPT_MFR SIMULATION TOOL

Adapt_MFR is a full radar simulation package that was designed and developed at DRDC Ottawa to model naval radars operating in a littoral environment. Support for both rotating and nonrotating phased array MFRs, as well as conventional rotating dishes such as volume search radars, is included. It incorporates models for land, sea, chaff, and rain clutter, as well as jammers. Adapt_MFR runs causally, producing detection output results for one beam at a time. Multiple waveforms and radar operational modes are available, including the dynamic and adaptive switching of waveforms. Adapt_MFR also includes the ability to model anomalous propagation, and to incorporate real terrain features through the importing of Digital Terrain Elevation Data (DTED) files.

An illustration of the high-level Adapt_MFR simulation architecture is presented in Figure 3.1. The framework consists of a series of modules (left-hand side) that describe the radar(s), target scenario, and environment which are required to provide input to the simulation. The simulation flow located in the center section of the figure represents the running code, which makes use of the data and associated functionality (algorithms, models, etc.). Adapt_MFR uses a tracker which employs an Interacting Multiple Model algorithm with a constant velocity model and a Singer maneuvering model for estimating target dynamics. The measurement models include range, range rate, bearing, and elevation. Detection-to-track data association is carried out using Nearest Neighbor (NN) JPDA [111].

As a result of the large parameter set and general versatility of the tool, there are many and varied modes in which it may be operated. There are, however, three basic modes of operation for Adapt_MFR, which are:

- calculator mode;
- simulation mode without tracker; and
- simulation mode with IMM tracker.

The calculator mode allows the user to compute preliminary detection results in a noncausal mode. The simulation modes are causal in nature and provide a complete simulation run, making the functionality of Adapt_MFR available to the user.

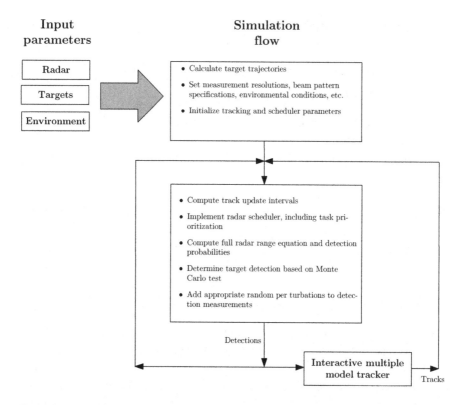

Figure 3.1 High-level overview of the simulation mode with IMM tracker in Adapt_MFR.

In order to analyze the performance of RRM techniques, Adapt_MFR is operated in the simulation mode with the IMM tracker. An overview of this mode is shown in Figure 3.1. To operate in this mode, user inputs are accepted through a graphical user interface and stored into corresponding radar, scheduling, environmental and other data structures. Target initial positions and trajectories are set by the user. The simulator runs in a loop, with time incremented in each pass by the dwell time of the radar beam, until the simulation time ends. Surveillance continues until a detection occurs and a confirmation is scheduled for that detection. For each successful target confirmation, a measurement report is sent to the tracker. Predictions are requested at specific scheduled times based on user-defined rules to determine track update intervals. Based on the radar scheduling algorithm being modeled, future surveillance and tracking beams are assigned at specific times. Adapt_MFR is capable of modeling networked radars with an arbitrary number of radars. Multiple-radar tracking is also enabled.

Adapt_MFR accurately assesses RRM performance by causally modeling radar operation on a beam-by-beam basis. Radar detections are input to the IMM tracker. The tracker is then capable of sending track update requests to the radar scheduler. Tracking performance is analyzed by comparing tracker outputs to ground truth data.

3.3 ADAPTIVE TECHNIQUES

Current and future defence systems will employ multifunction phased array radars to provide search, track, weapon control, and identification, all under software control. Existing RRM techniques utilize fixed sets of tasks and task priorities. Future RRM techniques are expected to employ adaptivity. The potential benefits of these adaptive control techniques and strategies are expected to include:

- optimization of the available radar timeline (search time vs. other tasks, etc.), leading to improved performance, especially when close to overload;
- adaptive modification of radars performance as the environment changes (e.g., ship moving from Open Ocean to Littoral, ECM);
- ability to reconfigure the radar for different applications through software control changes (e.g., from medium range air defence to BMD); and
- ability to rapidly reconfigure the radar to counter unforeseen threats, applications, and eventualities (e.g., introduction of a different threat into the scenario).

The objective of adaptive techniques is to develop approaches that optimize performance for an MFR in a dynamically changing environment. This section describes Adaptive RRM, which is characterized by three main techniques:

1. fuzzy logic prioritization;
2. time-balancing scheduling (TBS); and
3. adaptive update intervals for tracking.

These techniques are described here in more detail.

3.3.1 Fuzzy Logic Prioritization

Figure 3.2 shows the fuzzy logic decision tree and can be used to rank the relative importance of targets detected and tracked by the radar. This tool can be used to support task scheduling as well as allowing Adaptive RRM to

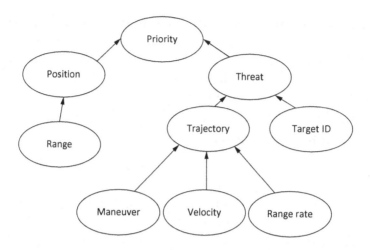

Figure 3.2 Decision tree for fuzzy logic adaptive target priority.

decide when track accuracy for low priority targets may be relaxed when the radar is overloaded. The ability of the radar to selectively relax the detection and tracking requirements for lower-priority radar tasks, coupled with the need to adapt the task priority order for task scheduling, provides a tool to support the implementation of graceful degradation of radar performance in extreme overload. Data inputs into the fuzzy logic tool such as track range, (radial) range rate, and velocity are generated by the track extractor. Target ID may be produced by a high-resolution radar (HRR) classification activity.

Figure 3.3 illustrates an example of the output from the fuzzy logic tool and how this is used to drive the use of tracking resources. The blue plot represents a target which has been classified as hostile while the red plot represents the same target trajectory when the target is identified as friendly. Clearly, the friendly target has been classified as less important and, therefore, Adaptive RRM uses this information to request less radar time for dedicated tracking of this target.

3.3.2 Time-Balancing Scheduling

Adaptive RRM uses the time-balancing algorithm for its beam-time alloca-tion [40]. Time balancing is a method that is often used in operating systems to allocate times dynamically for different processes. Figure 3.4 shows a time-balancing graph. A time-balance slope is defined for each beam type. For each time unit of 0.012007 s in this example, the simulation carries out the algorithm and determines which beam to execute. Once that beam is executed, the time balance is then stepped down. Two trials were performed

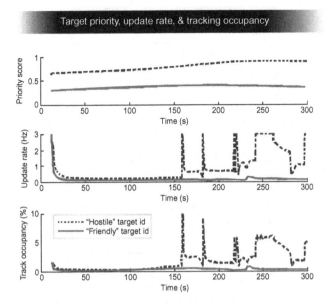

Figure 3.3 Illustration of the relative priority or importance of a target and the use of this information to drive the resource needs for dedicated tracking. The two examples (for friendly or hostile identification) are for the same target trajectory. Note that the transient spikes in update rates and occupancy are due to the tracker declaring a maneuver.

to verify the accuracy of this algorithm. After running the program, the values were geometrically verified using the time-balancing graph produced by the simulation. Proportions were created to convert between the number of pixels on the screen and the actual units in seconds. The results of the verification of the algorithm for Trials 1 and 2 are shown in Table 3.1. The algorithm was also graphically verified as shown in Figures 3.4 and 3.5, which illustrate the time-balancing algorithm graphs for Trials 1 and 2, respectively.

3.3.3 Adaptive Update Intervals for Tracking

This section describes the calculation of update intervals for Adaptive RRM. Assume a track's state estimate is:

$$X(k) = [x(k), \dot{x}(k), \ddot{x}(k), y(k), \dot{y}(k), \ddot{y}(k)], \tag{3.14}$$

at time t_k in the North-East coordinate system. The covariance matrix is $P(k)$. For simplicity, the coordinate z in the North-East-Down coordinate system is ignored. The azimuth position θ, velocity $\dot{\theta}$, and acceleration $\ddot{\theta}$ are calculated by a nonlinear function $[\theta, \dot{\theta}, \ddot{\theta}] = h(x, \dot{x}, \ddot{x}, y, \dot{y}, \ddot{y})$ which includes the following three equations:

Table 3.1 Time-Balancing Verification
Trial #1:
Actual time balance slopes
time_bal.slope_trk = 0.20;
time_bal.slope_det = 1;
Verification
Track Slope – $\dfrac{124 \text{ pixels}}{5 \cdot 0.012007 \text{ s}} = \dfrac{295 \text{ pixels}}{x \text{ seconds}}$ $m = \dfrac{124}{5 \cdot 0.012007 \cdot 295} \cdot \dfrac{72 \cdot 0.05}{123} = 0.19$ Detection Slope – $\dfrac{124 \textit{ pixels}}{5 \cdot 0.012007 \text{ s}} = \dfrac{24 \text{ pixels}}{x \text{ s}}$ $m = \dfrac{124}{5 \cdot 0.012007 \cdot 24} \cdot \dfrac{72 \cdot 0.05}{123} = 0.98$
Trial #2:
Actual time balance slopes
time_bal.slope_trk = 0.75;
time_bal.slope_det = 0.5;
Verification
Track Slope – $\dfrac{124 \text{ pixels}}{5 \cdot 0.012007 \text{ s}} = \dfrac{132 \text{ pixels}}{x \text{ s}}$ $m = \dfrac{124}{5 \cdot 0.012007 \cdot 132} \cdot \dfrac{72 \cdot 0.05}{123} = 0.76$ Detection Slope – $\dfrac{124 \text{ pixels}}{5 \cdot 0.012007 \text{ s}} = \dfrac{29 \text{ pixels}}{x \text{ s}}$ $m = \dfrac{124}{5 \cdot 0.012007 \cdot 29} \cdot \dfrac{72 \cdot 0.05}{123} = 0.51$

$$\theta = \tan^{-1}\left(\frac{y}{x}\right), \tag{3.15}$$

$$\dot{\theta} = \frac{\dot{y}}{r^2} - \frac{y\dot{x}}{r^2}, \tag{3.16}$$

$$\ddot{\theta} = \frac{x\ddot{y}}{r^2} - \frac{y\ddot{x}}{r^2}, \tag{3.17}$$

where $r = \sqrt{x^2 + y^2}$ is the horizontal range. With (3.15)–(3.17), the Jacobian matrix for the transformation $X \xrightarrow{\text{h}} [\theta, \dot{\theta}, \ddot{\theta}]$ is defined as follows:

$$H(k) = \frac{\delta h}{\delta X}\bigg|_{X(k)} = \begin{vmatrix} \frac{-y(k)}{r^2(k)} & 0 & 0 & \frac{x(k)}{r^2(k)} & 0 & 0 \\ \frac{y(k)}{r^2(k)} & \frac{-y(k)}{r^2(k)} & 0 & \frac{-x(k)}{r^2(k)} & \frac{x(k)}{r^2(k)} & 0 \\ \frac{y(k)}{r^2(k)} & \frac{-y(k)}{r^2(k)} & 0 & \frac{-x(k)}{r^2(k)} & \frac{x(k)}{r^2(k)} & 0 \end{vmatrix}. \tag{3.18}$$

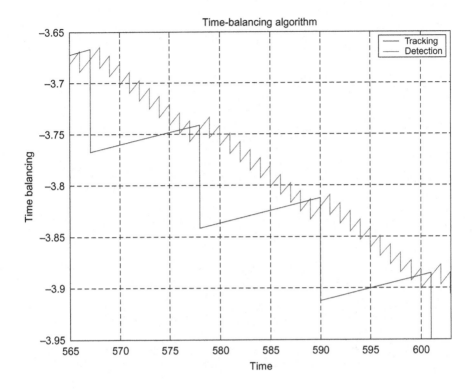

Figure 3.4 Time-balancing algorithm for Trial 1.

The covariance matrix $P_{az}(k)$ for the estimation error of $[\theta(k), \dot{\theta}(k), \ddot{\theta}(k)]$ is calculated as follows:

$$P_{az}(k) = H(k)P(k)H^{T}(k). \tag{3.19}$$

From $P_{az}(k)$, three quantities are available: the azimuth variance $a(k)$, azimuth/azimuth rate covariance $b(k)$, and azimuth rate variance $d(k)$. For medium priority targets, the time interval for the next updates is calculated by the following equation:

$$\tau^2(k) \leq \frac{E(k) - \sqrt{E^2(k) - 16A^2F(k)}}{8A^2}, \tag{3.20}$$

where

$$E(k) = 4Bhr(k)A + 16r^2(k)d(k), \tag{3.21}$$
$$F(k) = r2(k)B^2 - 16r^2(k)a(k) - 32\hat{\tau}(k)b(k). \tag{3.22}$$

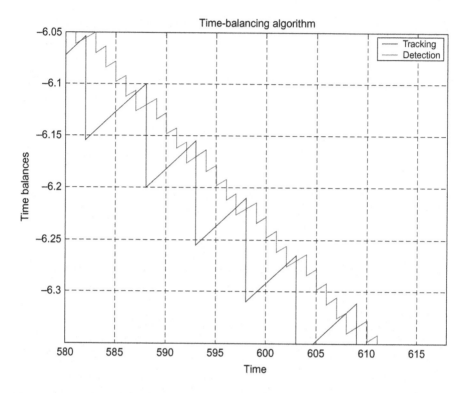

Figure 3.5 Time-balancing algorithm for Trial 2.

Similarly, for high priority targets, the equation is as follows:

$$\tau_h^2(k) \leq \frac{E_h(k) - \sqrt{E^2(k) - 16A^2 F_h(k)}}{8A^2}, \qquad (3.23)$$

where

$$E_h(k) = 4B_h r(k)A + 16r^2(k)d(k), \qquad (3.24)$$

$$F_h(k) = r2(k)B_h^2 - 16r^2(k)a(k) - 32\hat{\tau}_h(k)r^2(k)b(k), \qquad (3.25)$$

$$B_h = \frac{2B}{K}. \qquad (3.26)$$

In (3.20) and (3.23), $\hat{\tau}(k)$ and $\hat{\tau}_h(k)$ are approximations of $\tau(k)$ and $\tau_h(k)$. The purpose of the approximations is to simplify the calculation. Note that $\hat{\tau}_h(k)$ is a sensitive parameter and a smaller value is needed to generate feasible solutions. To avoid infeasible solutions, we can simply let it be zero, that is $\hat{\tau}_h(k) = 0$. $\hat{\tau}(k)$ can be the previous rate $\tau(k-1)$ for the same target.

The variable A is the maximum acceleration of the target (typically 1g or 2g in meters per second). The variable B is the beamwidth for the target direction. The accuracy factor K is a tuning parameter between 2 and 10, where 5 or 6 is reasonable value. When K is high ($K > 6$ for example), the calculation of $\tau_h(k)$ by (3.25) may become unstable. It is suggested that one checks the positiveness of the values that are to be square-rooted. In addition, a setup of minimum and maximum values for both $\tau(k)$ and $\tau_h(k)$ would also be helpful to avoid unrealistic update rates: between 2 and 4 s for medium-priority targets, and between 0.25 and 2 s for high-priority targets.

Note that an azimuth has to be converted to an angle relative to the boresight for the beamwidth calculation.

3.4 PERFORMANCE COMPARISON

For this comparison, the scenario length is 600 s. The radar transmits coherent pulsed waveforms with a peak power of 10 kW.

Two target scenarios are implemented in Adapt_MFR. Scenario 1 has a total of 52 targets, where each target is present in the scenario for some portion of the scenario timeline. Because all targets are not necessarily present at any given time, the total number of targets in the simulation may be fewer than 52 at any given time. The targets may be surface or airborne platforms. Each target is specified by a unique platform identification, location, velocity, trajectory, and RCS.

Scenario 2 has a total of 152 targets. These targets include the 52 targets from Scenario 1, 50 targets which are replicated from the original 52 targets, and 50 bird targets.

Nonadaptive RRM allocates 30% of the radar's timeline to tracking tasks, with the remainder allocated to surveillance tasks. There is no prioritization among the tracking tasks. As described in Section 3.3, Adaptive RRM utilizes fuzzy logic prioritization, time-balancing scheduling, and adaptive update rates for tracking. High-priority targets are tracked targets with priority greater than 0.7. These targets have a desired track update rate given by (3.23). Medium-priority targets have a priority value between 0.3 and 0.7, and have a desired track update rate given by (3.20). Low-priority targets have a priority value of less than 0.3. These targets are updated using track-while-scan; that is, there are no dedicated track update beams for low-priority targets.

To begin the analysis of this performance comparison, the behavior of a single target is considered. In Scenario 1, Target 1 is an airborne target that is in the field of regard at the start of the scenario and is initially approaching the radar. At approximately 150 s, the target starts to travel away from the radar and exits the field of regard at 300 s. The range and azimuth of Target 1 as a function of time are shown in Figure 3.6. The target track priority, as shown in Figure 3.7 is 0.85 until approximately 100 s, and then decreases slightly with time. Target 1 is high-priority target until 200 s and then becomes a medium-priority target, as shown by the requested track intervals in Figure 3.8. As a high-priority target, the requested track intervals are between 0.25 and 2 s, while as a medium-priority target, the requested track intervals are between 2 and 4 s.

Scenario 1 has a total of 52 targets, but not all targets are in the radar's field of regard at any given time. For Scenario 1, Figure 3.9a shows the total number of targets in the field of regard during the simulation. Also shown are the number of high-, medium-, and low-priority targets. The priority value, as calculated by the fuzzy logic prioritization technique, dynamically varies with the characteristics and dynamics of a target. As a result, the number of high-, medium-, and low-priority targets also varies throughout the simulation. For Scenario 2, Figure 3.9b shows the total number of targets, including the number of high-, medium-, and low-priority targets.

For Scenario 1, Figure 3.10 shows track completeness as a function of target index. Adaptive and Nonadaptive RRM have similar track completeness values for most targets, with some minor exceptions. Figure 3.11 presents track occupancy as a function of time. Examination of Figures 3.10 and 3.11 shows that Adaptive RRM has significantly lower track occupancy than Nonadaptive RRM, while achieving similar values of track completeness. The use of Adaptive RRM allows the radar to allocate less time to tracking while maintaining the same track completeness as Nonadaptive RRM.

Figure 3.12 shows the frame time for Scenario 1. Track confirmation beams are allocated to detection for the purpose of computing track occupancy. The scheduling of track confirmation beams increases the time until the first detection beam position is revisited, which results in increased frame time. In Figure 3.12, this is seen for Adaptive RRM between 150 and 200 s of the simulation when numerous confirmation beams are scheduled. The corresponding increase in frame time is evident.

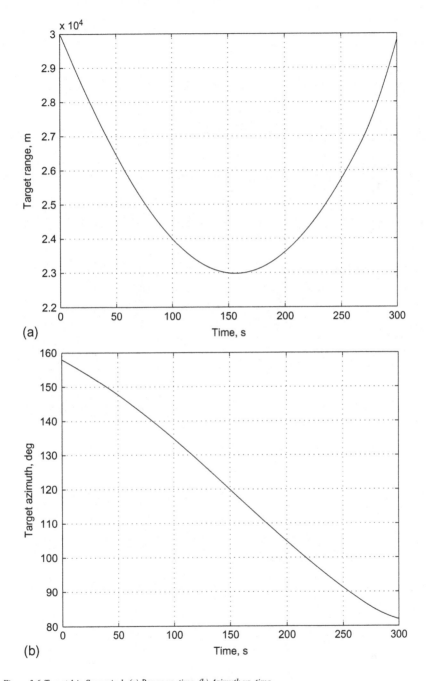

(a)

(b)

Figure 3.6 Target 1 in Scenario 1. (a) Range vs. time. (b) Azimuth vs. time.

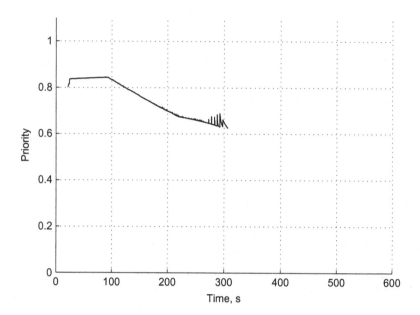

Figure 3.7 Track priority for Target 1 in Scenario 1.

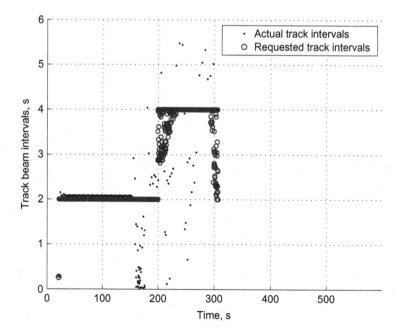

Figure 3.8 Track intervals for Target 1 in Scenario 1.

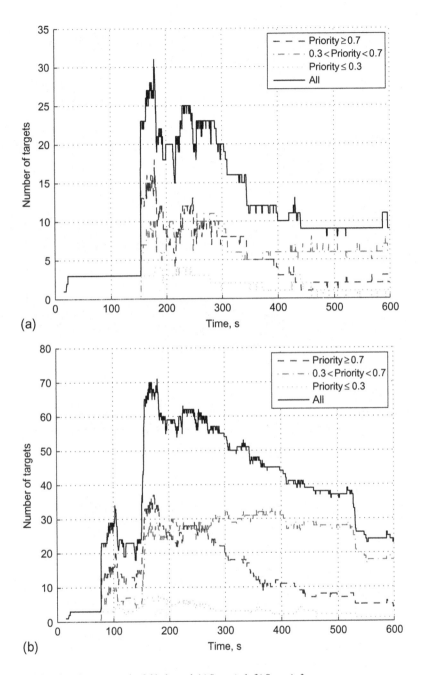

Figure 3.9 Number of targets in radar field of regard. (a) Scenario 1. (b) Scenario 2.

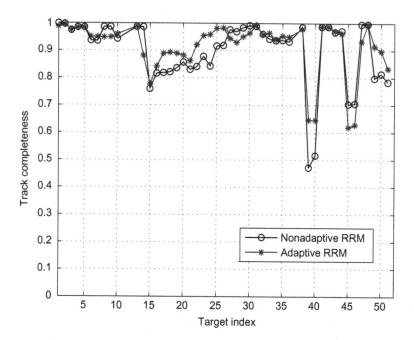

Figure 3.10 Track completeness for Scenario 1.

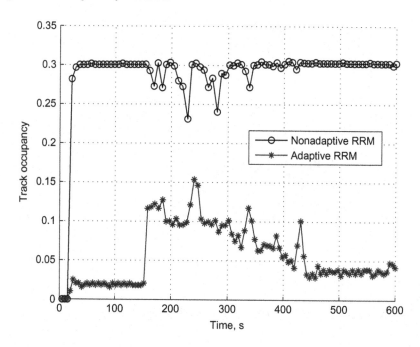

Figure 3.11 Track occupancy for Scenario 1.

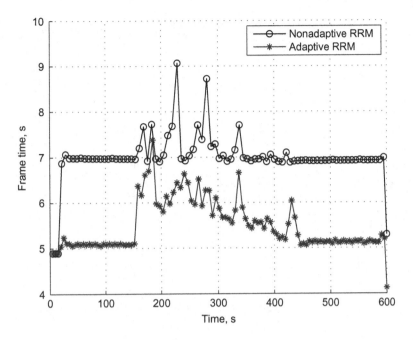

Figure 3.12 Frame time for Scenario 1.

Under Scenario 2, track completeness is shown in Figure 3.13. Adaptive RRM and Nonadaptive RRM have similar track completeness values. A notable exception is Target 7, for which Adaptive RRM has track completeness of 0.97, while Nonadaptive RRM has track completeness of 0.58. Targets 141 and 142, which are both birds, have higher track completeness for Nonadaptive RRM. Figure 3.14 presents track occupancy as a function of time. As was the case with Scenario 1, Scenario 2 results show that Adaptive RRM has significantly lower track occupancy and similar track completeness compared to Nonadaptive RRM. Figure 3.15 illustrates the frame time for Scenario 2.

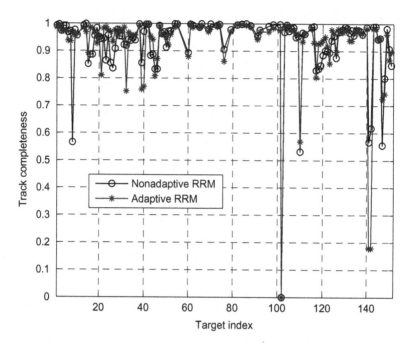

Figure 3.13 *Track completeness for Scenario 2.*

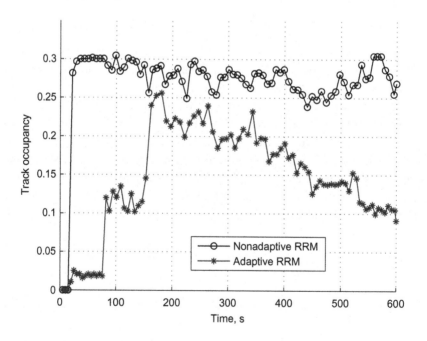

Figure 3.14 *Track occupancy for Scenario 2.*

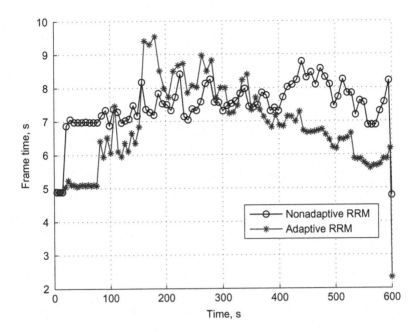

Figure 3.15 Frame time for Scenario 2.

Adaptive Scheduling Techniques

Adaptive Radar Resource Management. http://dx.doi.org/10.1016/B978-0-12-802902-2.00004-1
Copyright © 2015 Crown Copyright Published by Inc. All rights reserved.

This chapter presents two techniques for the adaptive scheduling of a multifunction radar (MFR). The optimal assignment scheduler (OAS) is described in Section 4.1. In Section 4.2, the two-slope benefit function scheduler (TSBFS) is presented.

4.1 OPTIMAL ASSIGNMENT SCHEDULER

4.1.1 Introduction

An MFR typically has two sets of priorities: function priorities and task priorities. The function priorities are predetermined by the radar mission. For example, the following priority levels exist:

1. high-priority tracks (highest priority);
2. track maintenance;
3. medium-priority tracks;
4. plot confirmation;
5. track initiation;
6. low-priority tracks;
7. surveillance and slow tracks; and
8. receiver calibration and built-in-test (BIT) (lowest priority).

Task scheduling can be interleaving and noninterleaving. In a few references, interleaving algorithms are proposed [112, 113], where there is an idle period between a pulse transmission and its receiving. The interleaving approach is not very practical. In this section, we use a model in which the sending and receiving sub-tasks are considered as one unique task.

Most radar scheduling algorithms fall into noninterleaving category. For example, Winter introduced a local search method to compute efficient schedules [114]. Cost functions for data link, tracking and searching are formulated. Linear programming is used to find the optimal schedule. The time-balancing scheduler (TBS) is a simple and efficient algorithm [115, 116].

In particular, the TBS was originally proposed and implemented in the MESAR system [115–117]. The scheduler keeps a time balance for each function. At any scheduling time, the radar picks the function with the maximum time balance for scheduling.

We propose a multi-level OAS, which is compared to the TBS. Accumulated scheduling delay and maximum delay are used as the performance measures. A simulation with over 400 detection beams and 40 targets is used to compare the two schedulers. In the simulation, each target requests a random update interval of 1-2 s.

4.1.2 Specifics of the Optimal Assignment Scheduler

Assume the radar has L functions and a time window $[t_k, t_{k+1}]$ is considered for scheduling. Within the scheduling time window, the radar is requested $[n_1, \ldots, n_L]$ beams by the L functions, respectively. A diagram of the scheduler is shown in Figure 4.1. The OAS includes four basic steps, which are specified as follows.

Step 1. Selection of beams: this process is mission strategic and determines how many beams will be scheduled. In an over-loading situation, some tasks will be discarded. For example, if the mission profile specifies the maximum tracking occupancy, the radar will be able to schedule a maximum number of tracking beams within a frame time.

Step 2. Formation of pseudo beams: this is created for all selected beams. The pseudo beams have time stamps, by simply using the detection dwell time in the proposed algorithm. No beam positions in azimuth and elevation are assigned in this step.

Step 3. Scheduling by OAS: this step finds the unassigned function with the highest priority so that the beams are assigned for a higher-priority function first. A cost matrix $[C_{ij}]$ can be built based on the task priority and time difference between the task time and the pseudo beam time:

$$[C_{ij}] = f(\Delta t_{ij}, p_i), \qquad (4.1)$$

where t_{ij} is the time difference between task i and pseudo beam j, and p_i is the priority of task i. The simplest cost function is the time difference where all tasks have the same priority:

$$[C_{ij}] = \Delta t_{ij}. \qquad (4.2)$$

If tasks have different priorities, the cost can be the product of the time difference and the priority value:

$$[C_{ij}] = \Delta t_{ij} * p_i. \qquad (4.3)$$

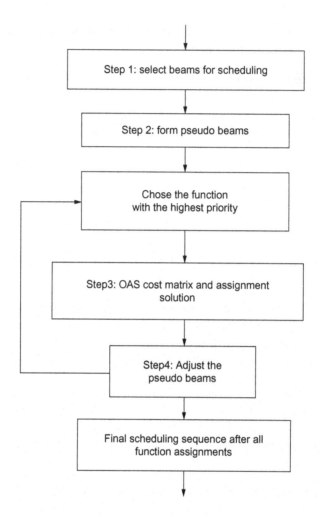

Figure 4.1 A diagram of OAS.

Step 4. Revise the pseudo beam list by replacing the assigned positions for the tasks in the function which is currently considered. Repeat step 3 until the last function, typically surveillance, which will be assigned to the unassigned beams.

4.1.3 Time-Balancing Scheduler

The TBS is a method that is often used in operating systems to allocate times dynamically for different processes. Figure 4.2 shows a time-balancing graph, where two functions have distinct time balance functions. The slope

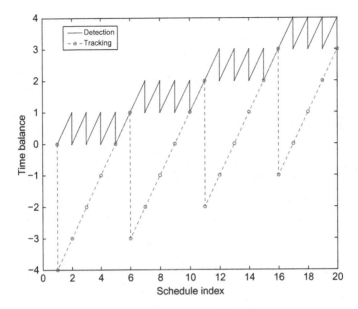

Figure 4.2 A TBS graph for 20% tracking occupancy.

and drop are the two parameters needed for the time balance of each functions. For convenience, the slope is set to one. By adjusting the slope values, a desirable tracking occupancy can be achieved. For example, in order to achieve 20% occupancy, the drops for the detection and tracking are one and four, which means there is one tracking beam after four detection beams. For example, a scheduling sequence is:

$$[1, 0, 0, 0, 0, 1, 0, 0, 0, 0, 1, 0, 0, 0, 0, 1, 0, 0, 0, 0],$$

where 1 stands for a tracking beam and 0 stands for a detection beam. The occupancies are 4/5, 1/5 for the two functions.

When more functions are involved, it becomes difficult to determine the occupancy. For instance, tracking 2 is prior to tracking 1, resulting in the following schedules:

$$[2, 1, 0, 0, 0, 0, 2, 1, 0, 0, 0, 0, 2, 1, 0, 0, 0, 0, 2, 1],$$

where 2 represents tracking 2. The occupancies are: 4/6, 1/6, 1/6 for the three functions.

The scenario is passed to a selection process, which determines how many tracking beams are to be scheduled. In the example in Figure 4.3,

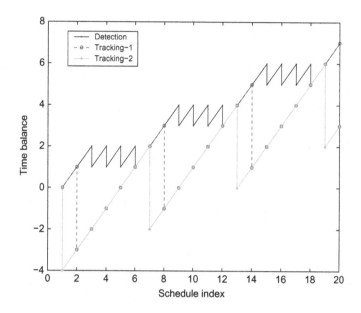

Figure 4.3 A TBS graph of three functions.

assuming the maximum track occupancy of 30%, all tracking beams will be scheduled. Note that the maximum tracking occupancy is a strategic decision, not an algorithmic decision. Once the maximum tracking occupancy is given, the maximum radar frame time can be calculated. Often, the maximum frame time is a surveillance requirement, which determines how fast a surveillance area can be revisited by the radar search beams.

4.1.4 Performance Evaluation
4.1.4.1 A Simulation Scenario
For simplicity, we consider a phased array radar with two functions: surveillance and tracking. The surveillance has a fixed number of 469 beams, each having a particular azimuth and elevation. There are 40 targets with confirmed tracks. Each track requests an update interval of 1-2 s. The dwell time for both surveillance and tracking is 0.01 s. The time window for assignment is [0, 6.25] s. Within this period of time, 156 track beams are requested. This requires a tracking occupancy of $156/(156+469) = 25\%$.

The scenario is passed to a selection process, which determines how many tracking beams are to be scheduled. Assuming the maximum track

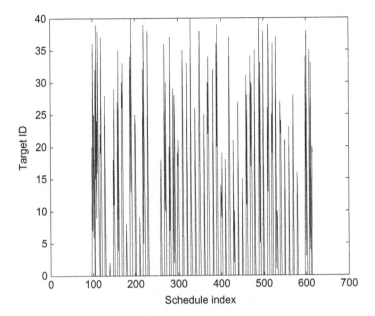

Figure 4.4 All targets and their requested scheduling times.

occupancy of 30%, all tracking beams will be scheduled. Note that the maximum tracking occupancy is a strategic decision, not an algorithmic decision. Once the maximum tracking occupancy is given, the maximum radar frame time can be calculated. Figures 4.4 and 4.5 show all the requested beams. It can be seen that multiple beams are required for the same time, resulting in many scheduling conflicts. For illustration purpose, the beam requests of the first ten targets are listed in Table 4.1.

4.1.4.2 Performance Comparison
The TBS and OAS are specified as follows:

1. The whole frame is used as the scheduling time window.
2. Tracking function has a higher priority over the detection function.
3. All tracking tasks have equal priority and the time difference is used as the cost function in OAS.
4. Both schedulers are designed for two-level assignment. First, tracking beams are assigned. For OAS, the assignment matrix is set up and the auction algorithm is used to find the best solution [13].
5. The TBS is modified to handle the same two-level assignment. Since only two functions are considered at each level, the occupancy can be exactly

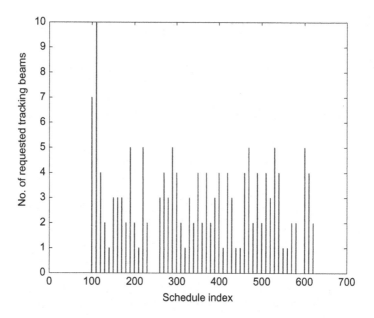

Figure 4.5 Requested 156 tracking beams.

Table 4.1 Beam Requests of the First Ten Targets				
Target	Update 1	Update 2	Update 3	Update 4
1	1	2.6	3.8	5.3
2	1.4	2.9	4.3	5.4
3	1.8	3.6	4.8	6
4	1.6	3	4.2	6.1
5	1.1	2.2	3.3	4.9
6	1	2.9	4.6	/
7	1	2.8	4.7	/
8	1.8	3.5	5	6.1
9	1.1	2.1	4	5.3
10	1.3	2.7	4.3	5.3

implemented. The tracking beams start at 1 s when the TBS starts. The drop values are 1 and 2.3654 for detection and tracking, respectively (see Figure 4.6). The value 2.3654 was calculated by 156/525, the ratio of the number of tracking beams and the number of total beams within the time balance window. All the tracking beams generated are assigned to the tracking beams based on the requested scheduling time, the earlier beams being scheduled first.

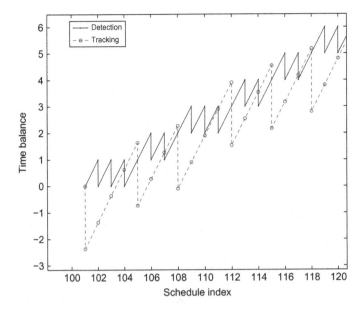

Figure 4.6 Part of the graph of the TBS.

For OAS, the assignment matrix is set up and the auction algorithm is used to find the best solution [118]. The testing scenario was processed by the two schedulers. The two performance measures used are the maximum delay and the accumulated delay. The results are shown in Table 4.2 and Figures 4.7 and 4.8. OAS has much better performance. Figure 4.9 compares the planned beams with 20% occupancy, the requested tracking beams and the actual scheduled tracking beams, where the requests with the same time are expanded near the request.

The TBS offers a uniform interval between tracking beams, which offsets the scheduling time from the requested time of tracking tasks. The OAS attempts to minimize the total offset—that is, the summation of all individual offsets.

In the simulation, we used one whole frame time to optimize the assignment. In practice, it is beneficial to have a much shorter scheduling

Table 4.2 Performance of Two Schedulers		
Schedulers	TBS (s)	OAS (s)
Maximum delay	0.49	0.06
Accumulated delay	30.08	1.62

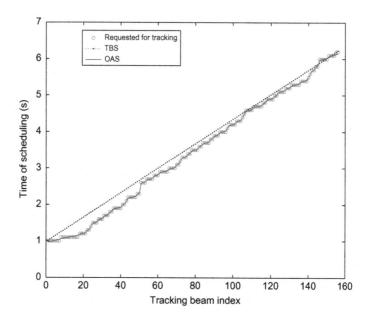

Figure 4.7 Comparison of scheduling time.

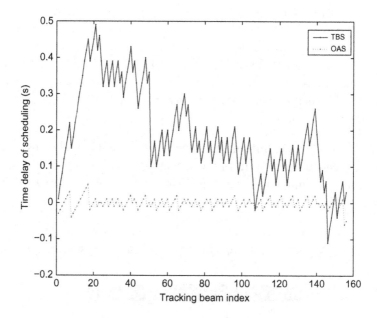

Figure 4.8 Comparison of maximum delays.

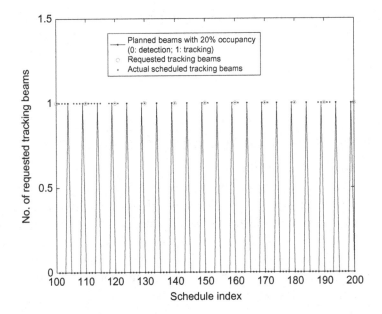

Figure 4.9 Scheduling results of OAS within 1-2 s.

time window, such as 0.5 to 1 s. With a shorter window, the radar is able to insert additional high-priority beams. In addition, the cost matrix has smaller dimensions, which significantly improves the efficiency of the optimization algorithm.

Also notice that we used the search dwell to form the initial pseudo beams. This formulation is accurate when the tracking waveform and search waveform have the same dwell time. If the dwell times are different, the formulation becomes an approximate one. The general formulation for the assignment cost matrix will be investigated in future work. We would also like to find out how the scheduling delays affect the final tracking performance, one of the ultimate goals of the radar resource management.

4.1.5 Summary

A new radar scheduler named OAS was proposed based on formulation of the radar scheduling into an optimal assignment problem. Using simulated radar data with detection and tracking tasks, the proposed OAS was tested and compared with an existing scheduler, the TBS. The new scheduler has much better performance as quantified in terms of the accumulated delay and maximum delay. In addition, the cost matrix formulation allows the use of priority information of each task within the same function, overcoming a drawback of other schedulers.

4.2 TWO-SLOPE BENEFIT FUNCTION SCHEDULER

4.2.1 Sequential Scheduler

The remainder of this chapter presents a technique called the sequential scheduler, which considers both look priorities and target dynamics in formulating a radar schedule. In overload situations, the sequential scheduler considers look priorities in deciding which looks to retain. For track updates, the error covariance varies with update time and target dynamics. The sequential scheduler accounts for individual target dynamics via the change in error covariance to schedule track updates.

The scheduler receives all look requests and formulates the schedule for a window of a fixed length of time. After the schedule has been carried out by the radar, the scheduler then formulates the schedule for the next window. Choosing shorter or longer windows have different advantages. If a shorter window is utilized, then newer look requests may be considered more quickly. Furthermore, the use of a shorter window results in a smaller number of look requests which must be processed at one time by the scheduler. If a longer window is utilized, a larger number of look requests must be processed, but the scheduling is carried out less frequently. The choice of window length is a balance between the need to consider new look requests rapidly and the desire to formulate schedules less frequently.

Tracking looks are by nature variable in quantity. At any given time, the radar maintains a variable number of tracks. Depending on the dynamics of the target, each track may need to be updated more frequently or less frequently. It is possible for the scheduler to receive no tracking look requests for a period of time. It is also possible for the scheduler to receive so many tracking look requests in a window that some requests must be dropped. Furthermore, a tracking look request is sensitive to its scheduled time. As the time between track updates increases, the uncertainty in the predicted position of the target increases. This results in the radar having to search a larger region to detect the target, which may increase the length of the radar look. If a long period of time elapses between track updates, then the track is lost. A long delay in scheduling a tracking look can result in the elimination of the associated task.

Surveillance looks consist of high-priority surveillance looks and lower-priority surveillance looks. High-priority surveillance looks are associated with short range pop-up targets which may pose an immediate, significant threat. These looks require a short dwell and a specified revisit interval. Lower-priority surveillance looks, which are associated with long-range search, may be thought of as a queue and are typically fixed in length, since each look request depends on the area to be monitored but not on the dynamics of a potential target. If the radar is not occupied with tracking looks or high-priority surveillance looks, then the next lower-priority surveillance look request in the queue should be scheduled. Each lower-priority surveillance look can be assigned a desired start time, as will be shown in Section 4.2.3.

Based on the differing priorities and nature of the looks considered, tracking looks and high-priority surveillance looks are called primary looks in this work. Lower-priority surveillance looks are called secondary looks. A method called the sequential scheduler is proposed. The scheduler, which is shown in Figure 4.10, consists of two components. The two-slope benefit function sub-scheduler generates a schedule of primary look requests. The gap-filling (GF) sub-scheduler considers the primary look schedule and

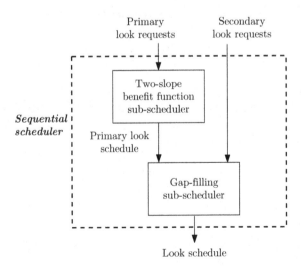

Figure 4.10 Illustration of the sequential scheduler.

schedules secondary look requests in any remaining idle intervals within the window. The TSBF sub-scheduler is described in Section 4.2.2. The GF sub-scheduler is described in Section 4.2.3. Scheduling primary look requests first ensures that tracking looks and high-priority surveillance looks are scheduled before lower-priority surveillance looks. The secondary looks are then scheduled to occupy as much of the radar window as possible.

This scheduler adaptively schedules look requests of arbitrary lengths, which are specified by the tracker or surveillance manager. Furthermore, the sequential scheduler is able to accommodate adaptive update rates, since desired start times may be chosen arbitrarily for both primary and secondary looks.

Note that the sequential scheduler does not place any constraints on the total time scheduled for tracking requests. It is possible to place a constraint on the sequential scheduler so that a maximum percentage of the window is devoted to tracking looks. Once the total length of scheduled tracking looks reached the maximum, surveillance looks would then be scheduled in the remaining radar time line. In this chapter, no such constraints are placed on the TSBF sub-scheduler.

4.2.2 Two-Slope Benefit Function Sub-Scheduler

This section describes the first component of the sequential scheduler, the two-slope benefit function sub-scheduler. The TSBF sub-scheduler receives primary look requests and each look request is either selected with a start time or dropped. Primary looks include tracking looks and high-priority surveillance looks, both of which have specified revisit intervals.

4.2.2.1 Preliminaries

Consider a time window $[T_1, T_2]$. Associated with this window, the sub-scheduler receives P primary look requests $L_1, L_2, \ldots, L_P$. Each look request L_n has the look parameters:

- l_n, the time required to complete the look, in seconds;
- t_n^*, the desired start time;
- s_n, the earliest start time;
- u_n, the latest start time;
- B_n^*, peak benefit;
- δ_n, slope for early scheduling; and
- Δ_n, slope for late scheduling.

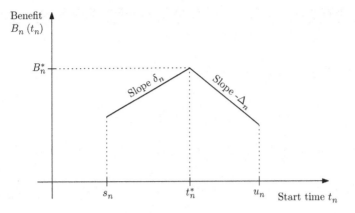

Figure 4.11 A two-slope benefit function.

The look parameters satisfy $s_n \leq t_n^* \leq u_n$. The interval $[s_n, u_n]$ is called the scheduling interval for L_n. All look requests are assumed to have a scheduling interval which lies within $[T_1, T_2]$. The slopes for early and late scheduling are restricted to $0 < \delta_n < \infty$ and $0 < \Delta_n < \infty$. If a look is scheduled, the start time t_n must satisfy $s_n \leq t_n \leq u_n$. The look request L_n is associated with a benefit function $B_n(t_n)$, which measures the benefit of selecting start time t_n. The benefit function is a two-slope function, which is given by:

$$B_n(t_n) = B_n^* + c_n(t_n - t_n^*), \tag{4.4}$$

where

$$c_n = \begin{cases} \delta_n, & s_n \leq t_n \leq t_n^* \\ -\Delta_n, & t_n^* < t_n \leq u_n \end{cases}. \tag{4.5}$$

A two-slope benefit function is illustrated in Figure 4.11.

Without loss of generality, the looks are ordered so that:

$$t_1^* \leq t_2^* \leq t_3^* \leq \cdots \leq t_P^*.$$

This ordering is a labeling convention and does not restrict the arbitrary choice of any of the look parameters. Note that it is not necessarily true that $s_n \leq s_{n+1}$ or $u_n \leq u_{n+1}$ for any $n = 1, \ldots, P - 1$, since the scheduling intervals for the looks may have different lengths.

Definition. A start time t_n for look L_n is a *viable start time* if $s_n \leq t_n \leq u_n$.

Table 4.3 List of Parameters for the TSBF Sub-Scheduler		
	Parameter	**Description**
Look	L_n	Look request
parameters	l_n	Time to complete look
(Section 4.2.2.1)	t_n^*	Desired start time
	s_n	Earliest start time
	u_n	Latest start time
	B_n^*	Peak benefit
	δ_n	Slope for early scheduling
	Δ_n	Slope for late scheduling
	t_n	Start time
	$B_n(t_n)$	Benefit function
Metrics	t_n'	Conditional earliest start time
calculation	E_n	Maximum delay within a sequence
(Section 4.2.2.5)	Q	Number of sequences
	D_q	Starting look for sequence q
	G_q	Maximum delay between sequences
Start time	α_n	Delay from earliest conditional start time
assignment	β_n	Allowable difference between delays
(Section 4.2.2.7)	$\overline{B}(t_1, \ldots, t_N)$	Total benefit
	v_n	Auxiliary variable for delay within a sequence
	w_n	Auxiliary variable for delay between sequences
	x_n	Additional variable for piecewise linear objective function
	$\hat{t}_n$	Optimal start time

If the start time for look L_n is t_n, then the look ends at time $t_n + l_n$. Because only one look may be executed at a time, the start time for the next scheduled look must be no earlier than $t_n + l_n$. This leads to the following definition.

Definition. Consider a set of looks $L_1, \ldots, L_P$ with ordered start times $t_1 < t_2 < t_3 < \cdots < t_P$. The set $L_1, \ldots, L_P$ is a *viable set* if:

$$t_n + l_n \leq t_{n+1}, \quad \text{for all } n = 1, 2, \ldots, P - 1.$$

Due to the large number of parameters used by the TSBF sub-scheduler, a list of parameters is specified in Table 4.3.

4.2.2.2 Selection of Look Parameters

The two-slope benefit function can be applied generally to primary look requests. In this section, a method for selecting the desired start time and slopes for early and late scheduling is presented. Tracking look requests are considered first.

To specify the desired start time, select a single coordinate of the tracking vector. The covariance of positional error is σ^2, which increases with the time between track updates. The desired start time $\hat{t}$ for the next track update is chosen so that updating the track at time $\hat{t}$ results in a specified error covariance σ_0^2. The update time value is calculated for the range, azimuth, and elevation coordinates, and the minimum of the calculated values is selected as the desired start time t^*.

Next, the slope for early scheduling is derived. If the start time for the track update is earlier than the desired start time, then the radar is increasing tracking occupancy and using radar resources that could be applied to other tasks. The benefit function is specified to reflect the amount of radar resources that are consumed by early scheduling of the track update. Therefore, the slope for early scheduling is inversely proportional to the look length l, that is:

$$\delta = \frac{1}{l}. \tag{4.6}$$

The slope for late scheduling is specified as follows. If the start time for the track update is later than the desired start time, then the error covariance increases. A linear estimate for the increase in error covariance is used, so that the error covariance of updated the track at time t is given by:

$$\sigma^2(t) \cong \sigma_0^2 + \zeta(t - t^*).$$

where ζ is the estimated slope. The benefit function is specified to reflect the increase in error covariance associated with scheduling a track update at a time later that the desired start time. The slope for late scheduling is inversely proportional to the increase in error covariance, so that:

$$\Delta = \frac{1}{\zeta}. \tag{4.7}$$

For a tracker using a Kalman filter with a constant velocity model, the estimated slope ζ is given by [119]:

$$\zeta = 2(\Omega_{PV} + \Omega_V t^*),$$

where Ω_{PV} is the position-velocity covariance and Ω_V is the velocity variance of the track at the last update. For a Kalman filter with an accelerating model, the estimated slope is given by:

$$\zeta = 2\Omega_{PV} + 2[2\Omega_{PA} + \Omega_V]t^* + 6\Omega_{VA}(t^*)^2 + 4\Omega_A(t^*)^3,$$

where Ω_{PA} is the position-acceleration covariance, Ω_{VA} is the velocity-acceleration covariance, and Ω_A is the acceleration variance of the track at the last update.

Next, consider high-priority surveillance looks. For each high-priority surveillance task, specify a linear cost function $\gamma(t) = \zeta t$ which measures the cost of waiting t seconds between successive looks. A desired cost γ_0 is also selected, and the desired start time is chosen so that $t^* = \gamma_0/\zeta$. The slopes for early and late scheduling are given by (4.6) and (4.7), respectively.

4.2.2.3 Radar Loading of Primary Looks Within the Time Window

For a set of primary look requests $L_1, \ldots, L_P$ with lengths $l_1, \ldots, l_P$, define:

$$\bar{l} = \sum_{n=1}^{P} l_n.$$

This is the minimum total time required to complete all looks. Also define:

$$\tau = \max_n(u_n + l_n) - \min_n s_n.$$

This is the maximum amount of radar time line available for the given set of looks. Note that in many cases, it may not be possible for all looks to be completed in the time τ, because this would require that multiple looks be executed by the radar simultaneously. The quantity τ is defined to facilitate a definition of radar loading.

Definition. The radar is *underloaded* during the time window if $\bar{l} \leq \tau$.

An underloaded radar may be able to schedule all P looks in τ seconds. However, it may be able to schedule all looks only by starting some of them at times other than the desired start times. Note that being underloaded is a necessary, but not sufficient, condition for the radar to execute all looks.

Definition. The radar is *overloaded* during the time window if $\bar{l} > \tau$.

An overloaded radar cannot schedule all P looks, and some looks must be dropped. With these definitions of loading, a set of looks $\{L_n\}$ can be categorized into one of three conditions, as follows.

Definition. A set of look requests $L_1, \ldots, L_P$ is in *Condition I* if the radar is underloaded and

$$t_n^* + l_n \leq t_{n+1}^*, \quad \text{for all } n = 1, 2, \ldots, P - 1. \tag{4.8}$$

That is, every look can be scheduled at its desired start time. This case is straightforward and does not require the sub-scheduler to make any decisions about shifting start times away from desired start times or dropping looks.

Definition. A set of look requests $L_1, \ldots, L_P$ is in *Condition II* if the radar is underloaded and (4.8) does not hold.

For a set of look requests in Condition II, it may be possible for all looks to be scheduled, but only if at least one of the looks does not start at its desired start time. It may also be the case that some looks must be dropped.

Definition. A set of look requests $L_1, \ldots, L_P$ is in *Condition III* if the radar is overloaded.

In this case, the radar is overloaded, and some looks must be dropped. The sub-scheduler must also select start times for the looks.

For Condition I, all looks can be scheduled at their desired start times, so scheduling is trivial. For Conditions II and III, the sub-scheduler may have to decide which looks to drop, and must schedule each of the looks that are retained. The TSBF sub-scheduler is used for look requests that are in Condition II or III.

4.2.2.4 Sub-Scheduler Overview
The input to the TSBF sub-scheduler is a set of P primary look requests. The output of the sub-scheduler is a viable subset of N looks, where $N \leq P$, and start times for each of the N looks. The viable subset may be the entire set of P look requests. The main components of the TSBF sub-scheduler are shown in Figure 4.12.

A radar scheduler must decide whether or not to schedule a look request, and if the look is to be scheduled, select a start time. The TSBF sub-scheduler carries out these two functions separately, by first deciding which looks are to be scheduled, and then selecting start times for the resulting subset.

Recall that the look requests are ordered so that $t_1^* \leq t_2^* \leq t_3^* \leq \cdots \leq t_P^*$. The TSBF sub-scheduler assumes that looks are scheduled in sequence, so that $t_i < t_j$ for $i < j$. This assumption significantly reduces the computational requirements of the sub-scheduler, but the resulting schedule is not necessarily optimal.

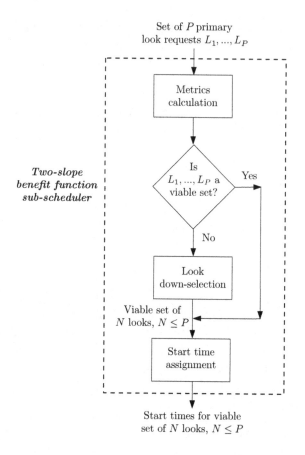

Figure 4.12 Overview of the two-slope benefit function sub-scheduler.

The set of P look requests $L_1, \ldots, L_P$ are first processed to produce a set of look metrics. Details of the metrics calculation algorithm are given later. The resulting metrics are used to determine if the look requests are viable. If the set of look requests is viable, then the viable set is sent to the start time assignment algorithm. If the set of look requests is not viable, then the set is subject to the look down-selection algorithm, which produces a viable subset of looks. This subset will necessarily be a strict subset, since one or more look requests will need to be dropped. The start time assignment algorithm produces a set of start times that maximize the total benefit of the viable set of looks.

Previous work has made the assumption that the local arrangement of start times has little effect on system performance [25]. However, when there is a large disparity between the error covariance slopes of tracked targets, start time adjustments may have a significant effect on error covariance. The TSBF sub-scheduler influences the earliness or lateness of looks through the use of benefit functions.

4.2.2.5 Metrics Calculation

The metrics calculation algorithm is the first stage of the TSBF sub-scheduler. This algorithm is also used within the look down-selection algorithm.

Let $L_1, \ldots, L_P$ be the input set of look requests. The looks are assumed to be scheduled in order so that $t_1 < t_2 < t_3 < \cdots < t_P$. A number of metrics are calculated, including $\{t'_n\}_{n=1}^{P}$, $\{E_n\}_{n=1}^{P}$, an integer Q where $1 \leq Q \leq N$, $\{D_q\}_{q=1}^{Q}$, and $\{G_q\}_{q=1}^{Q-1}$ when $Q \geq 2$. The metrics $\{t'_n\}_1^{P}$ are the earliest available start times for each look, given that each previous look has been scheduled at its earliest available start time. The metrics $\{E_n\}_{n=1}^{P}$, Q, $\{D_q\}_{q=1}^{Q}$, and $\{G_q\}_{q=1}^{Q-1}$ quantify the maximum delay that can be applied to each look while remaining viable.

The metrics $\{E_n\}_{n=1}^{P}$ will be utilized to determine whether the set of look requests is viable. For a viable set of looks, all of the metrics will be used to assign the start times.

The metrics are calculated as follows.

1. Let $t'_1 = s_1, E_1 = u_1 - s_1, q = 1, D_q = 1$, and $n = 2$.
2. Let $t'_n = \max(s_n, t'_{n-1} + l_{n-1})$ and $E_n = u_n - t'_n$. If $s_n > t'_{n-1} + l_{n-1}$, then let $G_q = s_n - t'_{n-1} - l_{n-1}, q = q + 1$, and $D_q = n$.
3. If $n = P$ then $Q = q$ and stop. Otherwise, let $n = n + 1$ and go to step 2.

The metrics $\{t'_n\}$ are a set of start times which satisfy $t'_n \geq s_n$ for all n. However, $\{t'_n\}$ do not necessarily satisfy $t'_n \leq u_n$ for all n. An interpretation for $\{t'_n\}$ is as follows. The start time t'_1 is the earliest possible start time for L_1. For this choice of start time, L_1 ends at time $t'_1 + l_1$. The start time t'_2 is the earliest possible start time for L_2 given that L_1 started at time t'_1. If $t'_1 + l_1 \geq s_2$, then L_2 can start at time $t'_1 + l_1$, but if $t'_1 + l_1 < s_2$, then L_2 must wait until time s_2 to start, since the start time must satisfy $t'_2 \geq s_2$. For L_n the start time t'_n is the earliest possible start time given that L_m started at time t'_m for $m = 1, \ldots, n - 1$. The start times $t'_1, \ldots, t'_P$ are chosen without considering the latest start times.

For each n, E_n is the difference between u_n and t'_n. The definition of $\{E_n\}_{n=1}^P$ results in a test for the viability of the set of look requests.

Test for viability: If $E_n \geq 0$ for $n = 1, \ldots, P$, then $\{L_1, \ldots, L_P\}$ is viable set and $\{t'_n\}_{n=1}^P$ is a set of viable start times.

To prove this, note that by definition $t'_n \geq s_n$ for $n = 1, \ldots, P$ and $t'_{n+1} \geq t'_n + l_n$ for $n = 1, \ldots P - 1$. If $E_n \geq 0$ for all n, then $u_n \geq t'_n$ for all n, which shows that $t'_1, \ldots, t'_P$ are viable start times. Therefore, $L_1, \ldots, L_P$ is a viable set.

The metrics calculation algorithm partitions the look requests into Q sequences. The first look request in sequence q is denoted D_q, and the first look of the first sequence is look L_1. For every look request in a sequence, except for the last look, the end time of the look request equals the start time of the next look request. For all sequences except the last sequence, the difference between the end time of the last look of sequence q and the start time of the next sequence is G_q. Because Q is the last sequence, G_Q is undefined.

Let $E_n < 0$ for a given n, and let q be the sequence containing look L_n. An explanation of the condition $E_n < 0$ is as follows. Since L_n is in sequence q, it is known that:

$$t'_n = s_{D_q} + \sum_{i=D_q}^{n-1} l_i,$$

and that $u_n < t'_n$. If all looks prior to L_n in sequence q are scheduled at their earliest possible time, then a viable start time for L_n cannot be selected. In order for L_n to be scheduled at a viable start time, one or more of the prior looks in sequence q must be dropped.

The metrics calculation algorithm serves two purposes in the TSBF subscheduler. First, calculation of the metrics $\{E_n\}$ leads to a test for the viability of the set of look requests. Second, the look metrics are used by the start time assignment algorithm to compute the start times that maximize total benefit. It is evident that the metrics calculation algorithm primarily carries out scalar operations.

4.2.2.6 Look Down-Selection

The look metrics lead to a test for the viability of a set of look requests: if $E_n \geq 0$ for all n, then the set is viable. If the set is not viable, then the

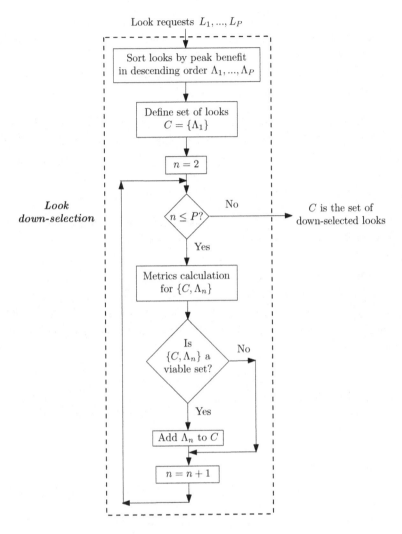

Figure 4.13 The look down-selection algorithm.

look down-selection algorithm drops one or more look requests to produce a viable set of looks.

A flow chart for the look down-selection algorithm is given in Figure 4.13. The look requests are sorted by peak benefit in descending order. The sorted look requests are labeled $\{\Lambda_n\}_{n=1}^{P}$. The goal of the down-selection is to select a viable subset C of $\{\Lambda_n\}_{n=1}^{P}$. Starting with the look request with the largest peak benefit, subsequent look requests are added one at a time to the set. The resulting new set is sent to the metrics calculation algorithm. If this new set is viable, then the most recent addition to the set is kept. Otherwise, the most recent addition is dropped. This process continues until all look requests have been considered. The look with the largest peak benefit, Λ_1, will always be included in C. Note that at least one look request will be dropped, because the original set of look requests is not viable.

As look requests are added to the set, the metrics are calculated to determine viability. If a set is not viable, dropping the most recent addition to the set is advantageous for two reasons. First, the most recent addition is the look with the smallest peak benefit that has been considered at that time. Second, dropping the most recent addition results in a viable set.

The relative peak benefit of a look request plays an important role in whether the look is down-selected. The look request with the largest peak benefit is always down-selected. To be down-selected, any other look request, together with all look requests with larger peak benefit, must be a viable set. For look requests with larger peak benefit, there are fewer other look requests in the set, which enhances the probability of being down-selected. Look requests with longer scheduling intervals also have a higher probability of being down-selected.

The look down-selection algorithm produces a viable subset C of the original set of look requests $L_1, \ldots, L_P$. The metrics calculation algorithm is executed $P - 1$ times by the look down-selection algorithm.

4.2.2.7 Start Time Assignment

The input to the start time assignment algorithm is a viable set of looks. If the original set of look requests was viable, then the input set is the original set of looks. If not, then the input set of looks is the viable subset generated by the look down-selection algorithm. The input set of looks is $\{L_n\}_{n=1}^{N}$. The metrics $\{t'_n\}_{n=1}^{N}$, $\{E_n\}_{n=1}^{N}$, Q, $\{D_q\}_{q=1}^{Q}$, and $\{G_q\}_{q=1}^{Q-1}$ (for $Q \geq 2$) were calculated for the input set of looks by the metrics calculation algorithm. The start time assignment algorithm schedules all looks in the input set.

Because all looks are scheduled, the start time t'_n is the earliest possible viable start time for L_n. All viable start times can be expressed as:

$$t_n = t'_n + \alpha_n, \quad n = 1, \ldots, N,$$

with the constraints:

$$0 \leq \alpha_n \leq E_n, \quad n = 1, \ldots, N, \tag{4.9}$$

$$\alpha_n \le \alpha_{n+1} + \beta_n, \quad n = 1, \ldots, N-1, \tag{4.10}$$

where

$$\beta_n = \begin{cases} 0, & \text{if } L_n \text{ and } L_{n+1} \text{ are in the same sequence} \\ G_q, & \text{if } L_n \text{ is in sequence } q \text{ and } L_{n+1} \text{ is in sequence } q+1 \end{cases}. \tag{4.11}$$

The start time assignment algorithm calculates viable start times to maximize the total benefit, which is given by:

$$\overline{B}(t_1, \ldots, t_N) = \sum_{n=1}^{N} B_n(t_n). \tag{4.12}$$

It will be shown that the benefit functions can be expressed in the form:

$$B_n(t_n) = B_n(t'_n) + f_n(\alpha_n). \tag{4.13}$$

Total benefit can then be expressed as

$$\overline{B}(t_1, \ldots, t_N) = \sum_{n=1}^{N} B_n(t_n) = \sum_{n=1}^{N} \left[B_n(t'_n) + f_n(\alpha_n) \right].$$

Therefore choosing viable start times which maximize (4.12) is equivalent to choosing $\{\alpha_n\}_1^N$ to maximize:

$$\sum_{n=1}^{N} f_n(\alpha_n), \tag{4.14}$$

subject to the linear inequality constraints (4.9) and (4.10). Furthermore, if $f_n(\alpha_n)$ is linear in α_n, then the maximization problem is a linear program and can be solved by applying the simplex method for linear programming [120].

It is shown that a benefit function can be expressed in the form given by (4.13). First, consider the case where $t'_n > t^*_n$:

$$\begin{aligned} B_n(t_n) &= B_n(t'_n + \alpha_n) \\ &= B_n^* - \Delta_n(\alpha_n - t^*_n + t'_n) \\ &= B_n(t'_n) - \Delta_n \alpha_n \end{aligned}$$

Therefore $f_n(\alpha_n)$ can be expressed as a linear equation:

$$f_n(\alpha_n) = -\Delta_n \alpha_n.$$

Next, consider the case where $t'_n \le t^*_n$. It is seen that:

$$B_n(t_n) = B_n(t'_n + \alpha_n)$$

$$= \begin{cases} B_n^* - \delta_n(t_n^* - t_n') + \delta_n \alpha_n, & 0 \le \alpha_n < t_n^* - t_n' \\ B_n^* - \delta_n(t_n^* - t_n') + \delta_n(t_n^* - t_n') - \Delta_n(\alpha_n - t_n^* + t_n'), & t_n^* - t_n' \le \alpha_n \le E_n \end{cases}$$
$$= B_n(t_n') + f_n(\alpha_n),$$

where the function $f_n(\alpha_n)$ is given by:

$$f_n(\alpha_n) = \begin{cases} \delta_n \alpha_n, & 0 \le \alpha_n < t_n^* - t_n' \\ \delta_n(t_n^* - t_n') - \Delta_n(\alpha_n - t_n^* + t_n'), & t_n^* - t_n' \le \alpha_n \le E_n \end{cases} . \quad (4.15)$$

In this case, $f_n(\alpha_n)$ is a piecewise-linear equation. The traditional simplex method for linear programming requires that the objective function be linear. Simplex methods for piecewise-linear objective functions have been developed [121, 122]. The approach taken here is to convert the piecewise-linear objective function to an equivalent linear objective function through the use of additional variables [123]. This then allows for the application of the traditional simplex method. Returning to (4.15), when $t_n^* - t_n' \le \alpha_n \le E_n$:

$$f_n(\alpha_n) = \delta_n(t_n^* - t_n') - \Delta_n(\alpha_n - t_n^* + t_n')$$
$$= \delta_n \alpha_n - (\delta_n + \Delta_n)(\alpha_n - t_n^* + t_n').$$

Therefore the function $f_n(\alpha_n)$ is given by:

$$f_n(\alpha_n) = \begin{cases} \delta_n \alpha_n, & 0 \le \alpha_n < t_n^* - t_n' \\ \delta_n \alpha_n - (\delta_n + \Delta_n)(\alpha_n - t_n^* + t_n'), & t_n^* - t_n' \le \alpha_n \le E_n \end{cases} .$$

For $0 \le \alpha_n \le E_n$, $f_n(\alpha_n)$ can be expressed as the linear equation:

$$f_n(\alpha_n) = \delta_n \alpha_n - (\delta_n + \Delta_n)\phi_n,$$

where

$$\phi_n = \begin{cases} 0, & 0 \le \alpha_n < t_n^* - t_n' \\ \alpha_n - t_n^* + t_n', & t_n^* - t_n' \le \alpha_n \le E_n \end{cases}$$
$$= \max(0, \alpha_n - t_n^* + t_n').$$

The optimization problem is thus a linear program, which is summarized as follows. Let N_E be the set of all n for which $t_n' \le t_n^*$. Choose $\{\alpha_n\}_1^N$ and $\{\phi_n\}_{n \in N_E}$ to maximize (4.14), where

$$f_n(\alpha_n) = \begin{cases} \delta_n \alpha_n - (\delta_n + \Delta_n)\phi_n, & \text{if } t_n' \le t_n^* \\ -\Delta_n \alpha_n, & \text{if } t_n' > t_n^* \end{cases}, \quad (4.16)$$

subject to the constraints:

$$v_n = E_n - \alpha_n, \quad n = 1, \dots, N, \quad (4.17)$$
$$w_n = \alpha_{n+1} - \alpha_n + \beta_n, \quad n = 1, \dots, N-1, \quad (4.18)$$
$$x_n = \phi_n - \alpha_n + t_n^* - t_n', \quad n \in N_E, \quad (4.19)$$

where β_n is given by (4.11) and $\alpha_n, \phi_n, v_n, w_n, x_n \geq 0$. The simplex method is used to compute the values for the variables $\alpha_n, \phi_n, v_n, w_n, x_n$ which maximize (4.14). Details of the simplex method are given in [124].

Let $\{\hat{\alpha}_n\}_1^N$ and $\{\hat{\phi}_n\}_{n \in N_E}$ be the set of variables which maximize (4.14). The start times that maximize the total benefit are then given by:

$$\hat{t}_n = t'_n + \hat{\alpha}_n, \quad n = 1, \ldots, N.$$

There are noteworthy special cases of this optimization problem. First, consider the case where $t'_n \geq t^*_n$ for all n. Examination of (4.16) shows that (4.14) is maximized by selecting $\alpha_n = 0$ for all n. The simplex method does not need to be carried out, and the start times $\{t'_n\}_{n=1}^N$ are optimal. Note that if $s_n = t^*_n$ for all n, then by definition $t'_n \geq t^*_n$ for all n.

Another special case occurs when $t^*_n = u_n$ for all n. In this case, the optimization problem simplifies to choosing $\{\alpha_n\}_1^N$ to maximize (4.14), where

$$f_n(\alpha_n) = \delta_n \alpha_n,$$

subject to the constraints:

$$v_n = E_n - \alpha_n, \quad n = 1, \ldots, N,$$
$$w_n = \alpha_{n+1} - \alpha_n + \beta_n, \quad n = 1, \ldots, N-1,$$

where β_n is given by (4.11) and $\alpha_n, v_n, w_n \geq 0$. In this case, the benefit function given by (4.4) is linear, as opposed to piecewise-linear for the general case. Therefore, the variables ϕ_n and x_n do not need to be introduced to convert the optimization problem to a linear program.

The benefit function was defined as a two-slope function, as given by (4.4) and (4.5). Benefit functions with other forms can also be formulated as shown in Section 4.2.7. The metric calculation and look down-selection algorithms are not dependent on the form of the benefit function. Only the optimization in the start time assignment algorithm is dependent on the form of the benefit function.

The TSBF sub-scheduler is applied to a fixed window. If a new primary look request is generated before the schedule is complete, then the scheduler must determine if there is an idle time interval at the desired start time for the new look request. If an idle time interval is not available, then a new schedule including all future look requests must be generated.

4.2.2.8 Impact of Look Parameters on Primary Look Schedule

For the TSBF sub-scheduler, each look request has a set of look parameters. The peak benefit affects whether the look request will be down-selected, and the slopes for early and late scheduling affect how close the start time will be to the desired start time.

If the original set of look requests is viable, then down-selection is not required. However, if the original set is not viable, then the peak benefit B_n^* plays an important role in the down-selection process. Because the look requests are ordered in descending order by peak benefit, look requests with larger peak benefits generally have a better chance of being down-selected. This is due to the fact that when it is being considered for inclusion in the set C, there are fewer other look requests, which together with the look request under consideration, may result in a set that is not viable. Previous work on task prioritization is a useful starting point for specifying the peak benefit. In particular, approaches based on fuzzy logic [12, 13] and neural networks [8] have produced methods for prioritizing target tracks.

The slopes for early and late scheduling represent the fluidity of the look. Once a set of viable look requests has been generated, the choice of slopes for early and late scheduling, δ_n and Δ_n, affect how close the look will be scheduled to its desired start time. If a look has larger values of δ_n and Δ_n relative to the other looks, then the look under consideration will be scheduled closer to its desired start time. This assumes that there is flexibility in choosing the start times for the viable set of looks. In cases where the set of looks is almost fully loaded, there may be little flexibility in choosing the start times.

The TSBF sub-scheduler also requires that the scheduling interval and desired start time be selected. Tracking update rates, which can be used to compute the desired start time have been studied in [21, 125, 126]. A formula for the latest start time has also been derived in [125]. The specification of the scheduling interval and its effect on tracking performance is an area that requires further study.

4.2.2.9 Summary

The TSBF sub-scheduler receives primary look requests and generates the start times for a viable subset of primary looks. If necessary, down-selection is carried out by a process which favors higher peak benefit look requests. Start times are chosen to maximize the total benefit of the viable look requests. It is shown that the maximization problem can be expressed as a linear program, which allows for the application of the simplex method.

4.2.3 Gap-Filling Sub-Scheduler for Secondary Looks

Inputs to the GF sub-scheduler are the primary look schedule generated by the TSBF sub-scheduler and a set of secondary look requests. If the entire window is occupied with primary looks, then the GF sub-scheduler does nothing, and the primary look schedule is the final schedule for the Sequential Scheduler. However, if there are any idle time intervals in the window, then the GF sub-scheduler attempts to schedule secondary looks in the idle time intervals. The secondary look requests are assumed to be organized as a queue.

4.2.3.1 Queue Management for Secondary Look Requests

The lower-priority surveillance function consists of M tasks, where each task involves the monitoring of a region in space which is defined by a distinct beam position or several beam positions of the radar. Each task generates a look request which is sent to the sub-scheduler. When the look from Task m is carried out, a new look request from that same task is generated. Therefore, the sub-scheduler is always in possession of a single look request from each task. Let the look request from Task m be labeled λ_m, where $m = 1, \ldots, M$. For this study, it is assumed that the length of each secondary look is d for all looks. In general, the dwell time for different looks may have different lengths. Unlike with a primary look request, there is no scheduling interval associated with a secondary look request. There is no earliest start time, because from the point of view of Task m, lower-priority surveillance of beam position m is of maximum benefit. There is also no latest start time, because even if a long period of time has elapsed since beam position m was last monitored, it is beneficial to the radar to schedule λ_m, because the surveillance of the region associated with Task m will enhance the surveillance picture.

The sub-scheduler is always in receipt of M secondary look requests. Associated with each look request is the elapsed time ϵ_m, which computes the time which has elapsed since the last time a look from Task m was carried out. Allow the requests to be organized in a queue, so that the sub-scheduler chooses the first look request in the queue. Ordering the look requests in the queue will determine how the look requests are scheduled. Two different cases are considered: one where all look requests have equal priority, and one where the look requests have different priorities.

4.2.3.2 Equal Priority Looks

When all M looks have equal priority, then the look requests form a first-in, first-out (FIFO) queue. After each look is chosen from the top of the queue and scheduled, it is inserted at the bottom of the queue. The FIFO queue is equivalent to ordering the look requests by ϵ_m in descending order. The look request with the largest ϵ_m is the next one to be scheduled.

Consider an example with six looks, where each look has equal priority. Let the current time be time 0. Table 4.4 shows the queue at time 0. The queue is ordered in descending order of elapsed time ϵ_m. If a secondary look is to be scheduled, look λ_3 will be chosen from the top of the queue. Now assume that a secondary look is not scheduled at time 0 nor at any time between time 0 and time t. Although the values of ϵ_m will all increase by t between time 0 and time t, the ordering of the looks in the queue will not change. For this equal priority example, regardless of when the next look is scheduled, λ_3 will be the next look chosen.

4.2.3.3 Looks with Unequal Priorities

Now consider the case where the looks have different priorities. This is suitable for varying surveillance tasks where the looks have desired start times. In this case, each look has a priority value which is specified by a time ω_m in seconds. When the sub-scheduler selects a look to be scheduled, all looks with the smallest value of ω_m and for which $\omega_m \leq \epsilon_m$ are placed at the top of the queue. If there is more than one such look, then the looks are ordered by ϵ_m in descending order. This is a generalization of the case where all looks have equal priority, since choosing $\omega_m = \infty$ for all m results in an FIFO queue. In this scheme, higher-priority looks will have a smaller value of ω_m, which results in look λ_m being scheduled more often.

Table 4.4 Look Queue with Equal Priority Looks, at Time 0	
Look	Elapsed Time ϵ_m (s)
λ_3	6.7
λ_4	6.5
λ_5	6.1
λ_6	6.0
λ_1	5.3
λ_2	4.5

Returning to the example with six looks, the elapsed times are assumed to be the same as in the previous example with equal priority looks. Let the current time be time 0. In this case, each of the looks is assigned a priority value, as shown in Table 4.5. Looks λ_4 and λ_6 have an elapsed time which is greater than their priority, so they are placed ahead of the other looks in the queue. Look λ_6 has a smaller elapsed time than look λ_4, but λ_6 has a smaller value of ω_m so it is placed ahead of λ_4 in the queue. Look λ_2 has a priority value of $\omega_2 = 4.0$, but its elapsed time is less than its priority, so it is not advanced in the queue. At time 0, if a secondary look is to be scheduled, look λ_6 will be chosen from the top of the queue. If a look is not scheduled at time 0, but is chosen between time 0 and time 0.5 s, then look λ_6 will still be scheduled, since it will be at the top of the queue.

Assume that no secondary looks are scheduled at time 0 or at any time between time 0 and time 0.5. At time 0.5, the elapsed times for all looks will have increased by 0.5. Since $\omega_2 = \epsilon_2$ and ω_2 is the smallest priority value, look λ_2 will move to the top of the queue, as shown in Table 4.6. Thus, it is seen that the ordering in the queue can change with time, depending on the values of ω_m in relation to the elapsed time of each of the looks.

Table 4.5 Look Queue with Unequal Priorities, at Time 0

Look	Priority ω_m (s)	Elapsed Time ϵ_m (s)
λ_6	5.5	6.0
λ_4	6.0	6.5
λ_3	∞	6.7
λ_5	∞	6.1
λ_1	∞	5.3
λ_2	5.0	4.5

Table 4.6 Look Queue with Unequal Priorities, at Time 0.5

Look	Priority ω_m (s)	Elapsed Time ϵ_m (s)
λ_2	5.0	5.0
λ_6	5.5	6.5
λ_4	6.0	7.0
λ_3	∞	7.2
λ_5	∞	6.6
λ_1	∞	5.8

4.2.3.4 Sub-Scheduler Operation

The GF sub-scheduler starts with the primary look schedule that was generated by the TSBF sub-scheduler and considers all intervals in the window during which no primary looks have been scheduled. The goal is to schedule as many secondary looks as possible in each idle interval. Let I be the length of an idle interval. If $I < d$, then no secondary looks are scheduled. Otherwise, k looks are scheduled in the interval, where k satisfies:

$$kd \le I < (k+1)d.$$

When one or more looks are to be scheduled, the GF sub-scheduler computes the state of the queue at that time, and the looks at the top of the queue are scheduled.

4.2.4 Scheduling Examples

The sequential scheduler is a general scheduling technique that can be applied to primary and secondary look requests with arbitrary parameters. In this section, several examples are considered that demonstrate the characteristics of the scheduler. In each example, the scheduler receives a number of surveillance and/or look requests and produces a schedule for these requests for a single window. Although only a single window is considered in these examples, an operational scheduler would schedule looks requests sequentially for consecutive windows. In this section, primary looks consist of tracking looks only, and secondary looks consist of surveillance looks only.

For all examples, the start time of the window is 0 ms. The nominal end time of the window is 300 ms. The scheduler will schedule all looks that start before the nominal end time. The actual end time of the window is when the last look is completed. There are 20 surveillance look requests, all with equal priority. The queue of surveillance look requests is given in descending order by $\lambda_1, \lambda_2, \ldots, \lambda_{20}$. The length of each surveillance look is $d = 15$ ms. In the examples presented, $\delta_n = \Delta_n$ for all tracking look requests; however, in general δ_n and Δ_n can be selected independently of each other. Recall that a list of parameters for the TSBF sub-scheduler is provided in Table 4.3.

Example 1 (No tracking looks). In this example, no tracking look requests are received by the scheduler. As a result, surveillance look requests are scheduled according to the order of the look requests in the queue. For look λ_i, the start time is $15(i-1)$ and the end time is $15i$. Surveillance looks occupy 100% of the window.

Table 4.7 Look Parameters and Start Times for Example 2

	Look Parameters							Start Time
Look	s_n	t_n^*	u_n	l_n	B_n^*	δ_n	Δ_n	$\hat{t}_n$
L_1	5	32	60	15	500	2	2	32
L_2	68	95	115	20	1000	1	1	95
L_3	197	220	240	15	200	1	1	220

Example 2 (Tracking looks in Condition I). In this example, the scheduler receives three tracking look requests, where the look parameters are shown in Table 4.7. It is evident that $t_1^* + l_1 < t_2^*$ and that $t_2^* + l_2 < t_3^*$. Therefore, the tracking looks are in Condition I and can be scheduled at their desired start times, as shown in the last column of Table 4.7.

Given a track schedule, the GF sub-scheduler then schedules as many surveillance looks as possible in the idle time intervals. The resulting schedule is shown in Table 4.8. The column labeled "function" indicates whether a time interval is idle or is occupied by a surveillance look or a tracking look. Of the 20 surveillance looks in the queue, 17 are scheduled during the window, so that surveillance look λ_{18} is at the top of the queue at the end of the window. Of the window length of 310 ms, 82.3% is occupied by surveillance looks, 16.1% by tracking looks, and 1.6% is idle.

For this example, the scheduling of the tracking looks is not affected by many of the look parameters, such as the scheduling interval $[s_n, u_n]$, the peak benefit B_n^* and the slopes for early and late scheduling, δ_n and Δ_n. Examples 3A, 3B, 4A, and 4B present cases where not all tracking looks can be scheduled at their desired start times. In these cases, the look parameters will determine which look requests are down-selected and when they are scheduled.

4.2.4.1 Examples 3A and 3B: Tracking Looks in Condition II

Examples 3A and 3B consider a set of tracking looks that are in Condition II. The two examples have look requests which differ in their slopes for early and late scheduling to highlight the characteristics of the TSBF sub-scheduler.

Example 3A. In this example, the scheduler receives ten tracking look requests, where the look parameters are shown in Table 4.9. Recall the definitions of radar loading and look request conditions from Section 4.2.2.3. For this set of looks, $\bar{l} = 195$ ms and $\tau = 255$ ms. Furthermore, it is not the

Table 4.8 Output of Sequential Scheduler for Example 2

Function	Look	Start Time (ms)	End Time (ms)
Surveillance	λ_1	0	15
Surveillance	λ_2	15	30
Idle	–	30	32
Tracking	L_1	32	47
Surveillance	λ_3	47	62
Surveillance	λ_4	62	77
Surveillance	λ_5	77	92
Idle	–	92	95
Tracking	L_2	95	115
Surveillance	λ_6	115	130
Surveillance	λ_7	130	145
Surveillance	λ_8	145	160
Surveillance	λ_9	160	175
Surveillance	λ_{10}	175	190
Surveillance	λ_{11}	190	205
Surveillance	λ_{12}	205	120
Tracking	L_3	220	235
Surveillance	λ_{13}	235	250
Surveillance	λ_{14}	250	265
Surveillance	λ_{15}	265	280
Surveillance	λ_{16}	280	295
Surveillance	λ_{17}	295	310

Table 4.9 Look Parameters, Look Metrics, and Start Times for Example 3A

Look	Look Parameters							Look Metrics		Start Time
	s_n	t_n^*	u_n	l_n	B_n^*	δ_n	Δ_n	t_n'	E_n	$\hat{t}_n$
L_1	5	25	40	15	700	5	5	5	35	18
L_2	15	38	56	15	200	1	1	20	36	33
L_3	21	48	68	20	1000	10	10	40	33	48
L_4	45	50	70	15	500	4	4	60	15	68
L_5	75	88	114	25	900	8	8	75	39	88
L_6	130	142	168	20	300	4	4	130	38	135
L_7	135	155	192	25	700	5	5	150	42	155
L_8	150	170	225	15	600	6	6	175	50	180
L_9	172	210	220	20	900	10	10	190	30	210
L_{10}	195	225	240	20	200	2	2	210	30	230

case that $t_n^* + l_n < t_{n+1}^*$ for $n = 1, \ldots, 9$. Therefore, the tracking looks are in Condition II.

The TSBF sub-scheduler begins by carrying out the metrics calculation for the set of looks requests. The look metrics $\{t_n'\}$ and $\{E_n\}$ are shown in Table 4.9. As detailed in Section 4.2.2.5, $\{E_n\}$ specify the maximum delay that can be applied to each look within a sequence. Because the parameters $\{E_n\}$ are non-negative for all n, the set of look requests is viable. Recall that the metrics calculation algorithm partitions the set of look requests into a number of sequences. In this case, the set of look requests was partitioned into two sequences; that is, $Q = 2$. The parameters D_q and G_q corresponding to the sequences are shown in Table 4.10. As specified in Section 4.2.2.5, G_q quantify the maximum delay that can be applied to looks which are in different sequences. It is seen that the first sequence includes the look requests L_1 through L_5, while the second sequence includes look requests L_6 through L_{10}. The parameter G_2 is undefined since the metrics calculation algorithm always leaves G_Q undefined.

The set of look requests is viable, so look down-selection is not required and the peak benefit values have no effect on the final schedule produced by the TSBF sub-scheduler. The next step for the TSBF sub-scheduler is the start time assignment algorithm. The start times $\hat{t}_n$ computed by the TSBF sub-scheduler are shown in the last column of Table 4.9. The start time assignment algorithm uses the metrics $\{t_n'\}$ as initial start times and increments the start times to maximize the total benefit. Figure 4.14 shows the total benefit as the start times are incremented by the simplex algorithm. In this example, 12 iterations of the simplex method are required to compute the start times that maximize total benefit. As explained in Section 4.2.6, if the entering basic variable has zero value, then an iteration of the simplex method will result in no change in the objective function. This accounts for the iterations in Figure 4.14 where the objective function remains unchanged.

Table 4.10 Sequence Parameters for Example 3A

Sequence (q)	D_q	G_q
1	1	30
2	6	-

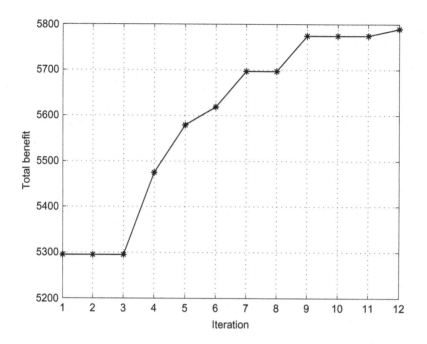

Figure 4.14 Increase in total benefit for the simplex method for Example 3A.

In this example, the two sequences of looks can be examined individually to understand how the total benefit is maximized. Consider the first sequence of looks, which consists of L_1 through L_5. For looks L_1, L_2, L_3, and L_5, $t_n' < t_n^*$ so that increasing the start time from t_n' increases the benefit function of these looks. However, for look L_4, $t_n' > t_n^*$ so that increasing the start time from t_n' decreases the benefit function of the fourth look. For each 1 ms increase to the start times, the net positive benefit is 20. Increasing the start times incrementally results in a net increase in total benefit until either one of the look start times reaches either its desired start time or its latest start time. In this case, this occurs when the start time for L_3 reaches 48 ms, which is its desired start time. Further increasing of the start times of the first four looks does not increase the total benefit. However, further increasing of the start time of the fifth look does increase the total benefit. The total benefit of the first sequence is maximized when the start time of the fifth look is the desired start time of 88 ms.

The second series of looks consists of looks L_6 through L_{10}. For each look except L_8, $t_n' < t_n^*$ so that increasing the start time from t_n' increases the individual benefit function. For look L_8, $t'_n > t^*_n$ so that increasing the start time from t'_n decreases the benefit function of this look. Increasing the start times for the second sequence increases the total benefit until the start time for L_7 reaches 155 ms, which is its desired start time. At this point, there is no net increase in benefit from increasing the start times for L_6, L_7, and L_8. However, increasing the start times for L_9 and L_{10} increases the net benefit until the start time for L_9 reaches 210 ms, which is its desired start time.

Because the radar is underloaded with tracking look requests, the sub-scheduler has flexibility in assigning start times for Example 3A. It is seen that the values of the slopes for early and late scheduling, δ_n and Δ_n, determine the start times which maximize the total benefit. The two looks with the largest slope values, L_3 and L_9, are both scheduled at their desired start times. Note that having larger slope values does not necessarily mean that a given look will be scheduled closer to its desired start time than a neighboring look with smaller slope values. In this example, look L_7 is scheduled at its desired start time while look L_8 is not, even though $\delta_7 = \Delta_7 = 5$ while $\delta_8 = \Delta_8 = 6$.

Given the look start times shown in the final column of Table 4.9, the GF sub-scheduler fills the idle time intervals with surveillance looks. Table 4.11 shows the output of the sequential scheduler for Example 3A. Seven surveillance looks are scheduled. Of the total window, 61.3% is occupied by tracking looks, 33.9% by surveillance looks, and 4.8% is idle.

The idle time intervals produced by the sequential scheduler are less than the length of the surveillance looks. Therefore, shorter surveillance look lengths result in shorter idle time intervals. The total amount of idle time produced by the schedule depends on both the length of the surveillance looks and on the total number of idle intervals generated by the TSBF sub-scheduler.

Example 3B. In Example 3B, the scheduler receives ten tracking look requests. The look parameters are identical to those from Example 3A, except that δ_2, Δ_2, δ_8, and Δ_8 are increased to 12. These modified values are indicated in bold italics in Table 4.12. The change in these parameters results in modified start times for seven of the ten looks. The new start times are indicated in bold italics in the last column of Table 4.12.

Due the increase in δ_2 and Δ_2, look L_2 is scheduled at 35 ms, which is 2 ms closer to its desired start time than in Example 3A. The neighboring

Table 4.11 Output of Sequential Scheduler for Example 3A

Function	Look	Start Time (ms)	End Time (ms)
Surveillance	λ_1	0	15
Idle	–	15	18
Tracking	L_1	18	33
Tracking	L_2	33	48
Tracking	L_3	48	68
Tracking	L_4	68	83
Idle	–	83	88
Tracking	L_5	88	113
Surveillance	λ_2	113	128
Idle	–	128	135
Tracking	L_6	135	155
Tracking	L_7	155	180
Tracking	L_8	180	195
Surveillance	λ_3	195	210
Tracking	L_9	210	230
Tracking	L_{10}	230	250
Surveillance	λ_4	250	265
Surveillance	λ_5	265	280
Surveillance	λ_6	280	295
Surveillance	λ_7	295	310

Table 4.12 Look Parameters, Look Metrics, and Start Times for Example 3B

Look	s_n	t_n^*	u_n	l_n	B_n^*	δ_n	Δ_n	t_n'	E_n	$\hat{t}_n$
				Look Parameters				Look Metrics		Start Time
L_1	5	25	40	15	700	5	5	5	35	*20*
L_2	15	38	56	15	200	*12*	*12*	20	36	*35*
L_3	21	48	68	20	1000	10	10	40	33	*50*
L_4	45	50	70	15	500	4	4	60	15	*70*
L_5	75	88	114	25	900	8	8	75	39	88
L_6	130	142	168	20	300	4	4	130	38	*130*
L_7	135	155	192	25	700	5	5	150	42	*150*
L_8	150	170	225	15	600	*12*	*12*	175	50	*175*
L_9	172	210	220	20	900	10	10	190	30	210
L_{10}	195	225	240	20	200	2	2	210	30	230

Note: Values in bold italics are variants from Example 3A.

looks L_3 and L_4 prevent L_2 from being scheduled at its desired start time of 33 ms. Look L_4 is scheduled at its latest start time of 70 ms, which forces L_3 to be scheduled to later than 50 ms and in turn forces L_2 to be scheduled no later than 35 ms.

Similarly, the increase in δ_8 and Δ_8 results in L_8 being scheduled closer to its desired start time. The neighboring looks L_6 and L_7 prevent L_8 from being scheduled at its desired start time of 170 ms. Looks L_6, L_7, and L_8 are scheduled at times t'_6, t'_7, and t'_8.

Examples 3A and 3B illustrate the effect of look parameters δ_n and Δ_n on the schedule computed by the TSBF sub-scheduler. Larger values of δ_n and Δ_n result in look L_n being scheduled closer to its desired start time.

4.2.4.2 Examples 4A and 4B: Tracking Looks in Condition III

Examples 4A and 4B consider a set of tracking looks that are in Condition III. The two examples have look requests with different values of peak benefit to highlight the characteristics of the TSBF sub-scheduler.

Example 4A. In this example, the scheduler receives 20 look requests, with look parameters as shown in Table 4.13. For this set of looks, $\bar{l} = 380$ ms and $\tau = 314$ ms, so that the set of looks is in Condition III. Therefore one or more look requests will need to be dropped to produce a viable set of looks.

The TSBF sub-scheduler begins by computing the look metrics for the set of look requests. These metrics which are calculated before down-selection are indicated in Table 4.13. The look requests are partitioned into one sequence; that is, $Q = 1$. It is seen that $E_n < 0$ for several looks. Therefore, the original set of look requests is not viable, and the sub-scheduler executes the look down-selection algorithm. As a result of the down-selection, looks L_7, L_{14}, L_{16}, and L_{18} are dropped to produce a viable set of looks. The look metrics for the viable set after down-selection are shown in Table 4.13. After down-selection, $E_{10} = 0$, which indicates that for look L_{10}, $t'_{10} = u_{10}$. Look must be scheduled at time $t'_{10} = 165$ ms. All down-selected looks are scheduled at the times $\{t'_n\}$ which were calculated after down-selection. Therefore, the slopes for early and late scheduling, δ_n and Δ_n, have no effect on the final schedule that is generated.

In the original set of look requests, there were two look requests, L_{14} and L_{16}, with a peak benefit of 100, which is the lowest value of peak benefit among all look requests. Both of these look requests were dropped in look

Table 4.13 Look Parameters, Look Metrics, and Start Times for Example 4A

Look	s_n	t_n^*	u_n	l_n	B_n^*	δ_n	Δ_n	t_n'	E_n	t_n'	E_n	$\hat{t}_n$
			Look Parameters					Look Metrics Before Down-Selection		Look Metrics After Down-Selection		Start Time
L_1	0	19	46	20	300	2	2	0	46	0	46	0
L_2	3	29	61	20	200	6	6	20	41	20	41	20
L_3	10	36	62	15	500	9	9	40	22	40	22	40
L_4	20	45	74	25	300	9	9	55	19	55	19	55
L_5	46	77	107	25	400	3	3	80	27	80	27	80
L_6	93	119	153	15	500	5	5	105	48	105	48	105
L_7	90	120	146	20	200	4	4	120	26	X	X	X
L_8	106	131	164	25	500	6	6	140	24	120	44	120
L_9	104	138	173	20	700	3	3	165	8	145	28	145
L_{10}	111	139	165	20	300	2	2	185	−20	165	0	165
L_{11}	144	175	208	15	800	5	5	205	3	185	23	185
L_{12}	164	196	226	15	200	3	3	220	6	200	26	200
L_{13}	168	202	233	20	500	1	1	235	−2	215	18	215
L_{14}	172	206	238	20	100	10	10	255	−17	X	X	X
L_{15}	180	213	243	20	300	5	5	275	−32	235	8	235
L_{16}	192	224	253	15	100	10	10	295	−42	X	X	X
L_{17}	196	227	261	20	300	8	8	310	−49	255	6	255
L_{18}	214	240	269	15	200	1	1	330	−61	X	X	X
L_{19}	219	248	281	20	300	7	7	345	−64	275	6	275
L_{20}	241	270	299	15	300	7	7	365	−66	295	4	295

Note: The symbol "X" indicates a look request that was dropped during down-selection.

down-selection. Of the four look requests with a peak benefit of 200, two look requests were dropped. Therefore, the four look requests which were dropped had the lowest peak benefit values. This will not necessarily be the case for all sets of look requests, as look down-selection process is also affected the length of the scheduling intervals and the relationships between scheduling intervals among the different look requests.

The schedule produced by the TSBF sub-scheduler has no idle intervals. Therefore, the tracking schedule is the final schedule, and no surveillance looks are scheduled in this example.

Example 4B. In this example, the scheduler receives 20 look requests. The look parameters are the same as in Example 4A, with the exceptions that

Table 4.14 Look Parameters, Look Metrics, and Start Times for Example 4B

Look			Look Parameters					Look Metrics Before Down-Selection		Look Metrics After Down-Selection		Start Time
Look	s_n	t_n^*	u_n	l_n	B_n^*	δ_n	Δ_n	t_n'	E_n	t_n'	E_n	$\hat{t}_n$
L_1	0	19	46	20	300	2	2	0	46	0	46	10
L_2	3	29	61	20	200	6	6	20	41	X	X	*X*
L_3	10	36	62	15	500	9	9	40	22	20	42	30
L_4	20	45	74	25	*1000*	9	9	55	19	35	39	45
L_5	46	77	107	25	400	3	3	80	27	60	47	70
L_6	93	119	153	15	500	5	5	105	48	93	60	95
L_7	90	120	146	20	200	4	4	120	26	108	38	*110*
L_8	106	131	164	25	500	6	6	140	24	128	36	130
L_9	104	138	173	20	*400*	3	3	165	8	153	20	155
L_{10}	111	139	165	20	300	2	2	185	−20	X	X	*X*
L_{11}	144	175	208	15	800	5	5	205	3	173	35	175
L_{12}	164	196	226	15	200	3	3	220	6	X	X	*X*
L_{13}	168	202	233	20	500	1	1	235	−2	188	45	190
L_{14}	172	206	238	20	*1000*	10	10	255	−17	208	30	*210*
L_{15}	180	213	243	20	300	5	5	275	−32	228	15	230
L_{16}	192	224	253	15	100	10	10	295	−42	X	X	X
L_{17}	196	227	261	20	300	8	8	310	−49	248	13	250
L_{18}	214	240	269	15	200	1	1	330	−61	X	X	X
L_{19}	219	248	281	20	300	7	7	345	−64	268	13	270
L_{20}	241	270	299	15	300	7	7	365	−66	288	11	290

Notes: The symbol "X" indicates a look request that was dropped during down-selection. Peak benefit values which differ from Example 4A are shown in bold italics. Start time values are in bold italics if the outcome of look down-selection differed from that in Example 4A.

the peak benefit for look L_9 is 400, and the peak benefits for looks L_4 and L_{14} are 1000. These modified values for peak benefit are indicated in bold italics in Table 4.14, which also presents the look metrics before and after down-selection and the start times $\hat{t}_n$ generated by the TSBF sub-scheduler.

The last column in Table 4.14 shows the start time for the look requests, with the symbol "X" indicating that the look was dropped. The start time entry is shown in bold italics if the outcome of the Look Down-Selection Algorithm was different from that in Example 4A. Looks L_2, L_{10}, and L_{12} were down-selected in Example 4A but dropped in Example 4B. Looks L_7 and L_{14} were dropped in Example 4A but down-selected in Example 4B.

Increasing the peak benefit of look L_{14} changed its outcome in the look down-selection algorithm. For all the other looks whose outcome of the look down-selection algorithm changed, their peak benefits were relatively low and unchanged from Example 4A. The changed peak benefit values of the other look requests caused the change in outcome of the look down-selection algorithm.

4.2.4.3 Summary
The examples presented in this section illustrate the schedules produced by the sequential scheduler for varying numbers of tracking look requests. When there are no tracking look requests or when all tracking look requests can be scheduled at their desired times, generating the tracking look schedule is trivial, as in Examples 1 and 2. However, when not all tracking look requests can be scheduled at their desired times, the TSBF sub-scheduler down-selects a viable set of looks if required, and schedules the viable set to maximize total benefit. Examples 3A and 3B showed that choosing larger values of δ_n and Δ_n for look L_n resulted in L_n being scheduled closer to its desired start time. Examples 4A and 4B showed that looks with larger values of peak benefit are more likely to survive look down-selection.

4.2.5 Comparison with Orman Scheduler
In this section, a scheduling scenario is considered, and the properties of a schedule generated by the sequential scheduler are compared to the properties of a schedule generated by the Orman scheduler [113]. The example is chosen to be similar to the overload radar example in [127, pp. 421-422].

Tracking and surveillance is carried out over a 25-s interval. Primary looks consist of tracking looks only. Thirty targets are detected and tracked, with track initialization for each target occurring randomly during the period from 5 to 20 s after the start of the interval. The parameters used for tracking are given in Table 4.15. For track updates, a probability of detection of one was used. In this example, tracking looks have higher priority than surveillance looks, which have a dwell time of 2 ms.

Table 4.15 Parameters for Tracked Targets

Number of Targets	Waveform Dwell Time (ms)	Update Interval (ms)
30	5	150

Schedules were generated concurrently for 1-s intervals. For a given interval, if a track was initiated during the interval, desired tracking looks were produced at 150 ms intervals after the track initiation. For existing tracks, desired tracking looks were produced at 150 ms intervals starting from the time of the last track update. For the sequential scheduler, each tracked target requires the specification of look parameters, and the slopes for early and late scheduling are given in Table 4.16. The slopes were chosen to illustrate scheduler properties.

Figure 4.15 shows the resource allocation over time for the sequential scheduler and the Orman scheduler. It is evident that both schedulers have similar performance with regard to allocation between tracking and surveillance looks. In the interval between 0 and 5 s, all resources are allocated to surveillance looks. As target tracks are initiated during the period between 5 and 20 s, more resources are allocated to tracking looks. In this interval, total occupancy is slightly less than 1. This is due to the structure of both schedulers, which allow short idle times in the radar schedule even when the radar is overloaded. Finally, in the interval between 20 and 25 s, all resources are allocated to tracking looks.

A total of 1820 tracking looks are scheduled. For all tracking looks, scheduled start times are compared to desired start times in Figure 4.16. Histogram plots of earliness or lateness with respect to desired start times are shown. Earliness values of equal to or less than −20 ms are grouped into a single bin at −20 ms, and lateness values of greater than or equal to 20 ms are grouped into a single bin at 20 ms. The sequential scheduler scheduled 43% of tracking looks at their desired start times, and all looks were scheduled within 13 ms of their desired start times. The Orman scheduler scheduled 56% of tracking looks at their desired start times, but 7% of tracking looks were scheduled 20 ms or more from their desired start times. Thus the Orman scheduler schedules more looks at their desired start time, but at the expense of scheduling other looks further from their desired start time, compared with the sequential scheduler.

Table 4.16 Slopes for Early and Late Scheduling for Tracked Targets		
Category	Number of Targets in This Category	Slope Values
High slope target	10	$\delta = \Delta = 35$
Medium slope target	10	$\delta = \Delta = 10$
Low slope target	10	$\delta = \Delta = 3$

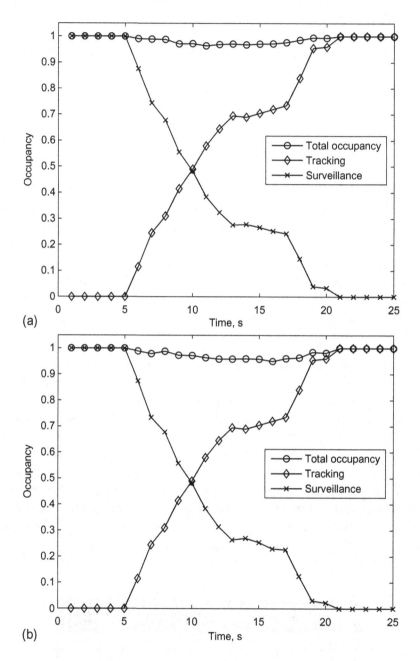

Figure 4.15 Achieved occupancy for the scheduling example. (a) Sequential scheduler. (b) Orman scheduler.

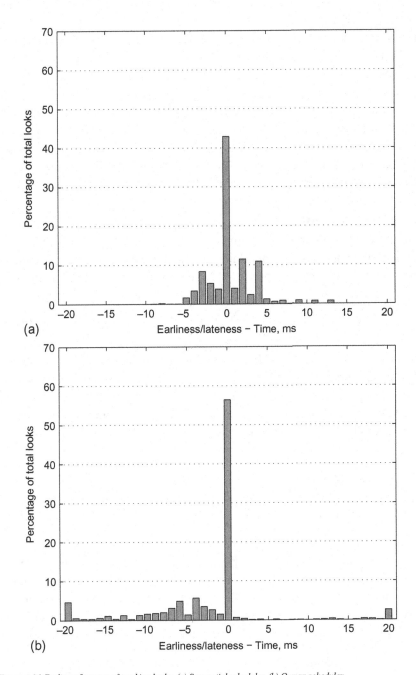

Figure 4.16 Earliness/lateness of tracking looks. (a) Sequential scheduler. (b) Orman scheduler.

Although Figure 4.16a shows earliness or lateness for all looks for the sequential scheduler, further insight into scheduler performance is gained by analyzing looks with different slopes. Figure 4.17a shows earliness/lateness for high slope looks, as specified in Table 4.16, while Figure 4.17b shows earliness/lateness for medium and low slope looks. For high slope looks, 62% of looks are scheduled at their desired start times, while only 32% of medium and low slope looks are scheduled at their desired start times. In an overload situation, it is not possible to guarantee that a look will be scheduled at its desired start time. However, this example illustrates that selecting a high slope value greatly enhances the likelihood that the scheduled start time will match the desired start time.

4.2.6 Simplex Method

This section shows how the simplex method for linear programming is applied to the optimization problem presented in Section 4.2.2.7. The linear program is as follows. Choose $\{\alpha_n\}_1^N$ to maximize the objective function:

$$\sum_{n=1}^{N} f_n(\alpha_n), \tag{4.20}$$

where $f_n(\alpha_n)$ is given by (4.16), subject to the constraints given in (4.17)–(4.19) and with the restrictions $\alpha_n, \phi_n, v_n, w_n, x_n \geq 0$. A feasible solution is a set of non-negative variables $\alpha_n, \phi_n, v_n, w_n, x_n$ that satisfy (4.17)–(4.19). The simplex method begins with an initial feasible solution and iteratively selects feasible solutions that increase the value of the objective function.

In this case, an initial feasible solution is given by $\alpha_n = 0$, for all n and $\phi_n = 0$ for all $n \in N_E$. Since $v_n = E_n - \alpha_n \geq 0$ for all n and $x_n = \phi_n - \alpha_n + t_n^* - t_n' \geq 0$ for all $n \in N_E$, it is evident that the variables α_n and ϕ_n are bounded, which shows that (4.20) is bounded and that a maximum exists.

To carry out the simplex method, variables are grouped into a set of nonbasic variables, which are set to zero, and a set of basic variables. For the initial feasible solution, the variables $\{\alpha_n\}_{n=1}^N$ and $\{\phi_n\}_{n \in N_E}$ are the nonbasic variables and the variables $\{v_n\}_{n=1}^N$, $\{w_n\}_{n=1}^N$ and $\{x_n\}_{n \in N_E}$ are the basic variables. The simplex method then iterates on the following two steps.

1. Determine the entering basic variable by computing which nonbasic variable, if allowed to take on a positive value, will increase the value of the objective function most rapidly.

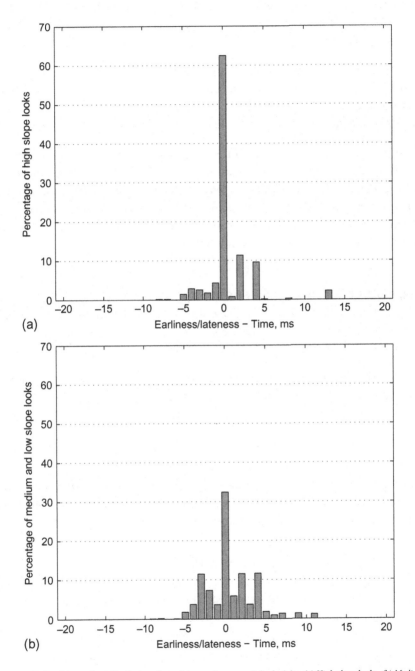

Figure 4.17 Breakdown of tracking look earliness/lateness for sequential scheduler. (a) High slope looks. (b) Medium and low slope looks.

2. Increase the value of the entering basic variable until one of the basic variables is forced to a value of zero. This basic variable is called the exiting basic variable.

The entering basic variable becomes a basic variable and the exiting basic variable becomes a nonbasic variable. Steps 1 and 2 are then repeated. The procedure stops when no entering basic variables exist, and the resulting values of the variables maximize the objective function.

If more than one variable qualifies as the entering basic variable, then the entering basic variable is chosen arbitrarily from among these candidates. If a basic variable has zero value, it may be necessary to exchange to choose this variable as the entering basic variable in order to proceed with the iteration. Such an operation will result in the objective function remaining constant for an iteration. This is illustrated in Figure 4.14.

4.2.7 Other Forms of the Benefit Function

For the TSBF sub-scheduler, the benefit function was defined as a two-slope function, as given by (4.4) and (4.5). This section examines other forms of the benefit function which may be considered.

The benefit function can be specified as an order-r function, which is defined by:

$$B_n(t_n) = B_n^* - c_n \|t_n - t_n^*\|^r, \tag{4.21}$$

where $c_n \in \Re$, $c_n \geq 0$, and $r \geq 2$ is an integer. Order-r benefit functions with $r = 2$, $r = 3$, and $r = 4$ are shown in Figure 4.18, when $c_n = 1$. It is seen that the benefit function decreases more slowly in $\|t_n - t_n^*\|$ as r increases. Given a viable set of N looks with order-r benefit functions, the start time assignment algorithm seeks to choose $\{\alpha_n\}_1^N$ to minimize:

$$\sum_{n=1}^{N} c_n \|\alpha_n - (t_n^* - t_n')\|^r, \tag{4.22}$$

subject to the constraints (4.9) and (4.10). This optimization can be carried out using Lagrange multipliers and solving for the Kuhn-Tucker conditions for optimality [128]. Solving this nonlinear program requires significantly more computation complexity than that of the simplex method.

Piecewise-linear benefit functions with multiple slopes may also be considered. These functions are more general than two-slope functions, while

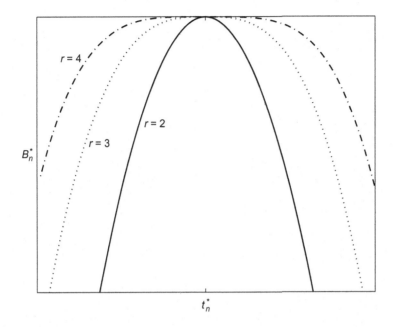

Figure 4.18 Order-r benefit functions for $c_n = 1$.

the process for selecting start times is still a linear program. However, in order to apply the simplex method, additional variables must be introduced to the problem. In general, one additional variable will need to be introduced for each additional linear piece of each benefit function [121].

CHAPTER 5

Radar Resource Management for Networked Radars

5.1 INTRODUCTION

Military systems are increasingly considering task force operation, where multiple platforms are deployed to an area of interest. This focus has resulted in research activity in sensor resource management, which optimizes the assignment of multiple sensors to multiple tasks [129]. Sensor resource management takes place at the Command and Control (C2) level and attempts to answer the question of what tasks should be assigned to various sensors. At the sensor (radar) level, radar resource management (RRM) considers the prioritization and scheduling of multiple tasks. This chapter

Adaptive Radar Resource Management. http://dx.doi.org/10.1016/B978-0-12-802902-2.00005-3

addresses the important, open problem of the coordination of RRM among networked radars.

Track scheduling for networked radars has been considered by He and Chong [130, 131], who model the sensor scheduling problem as a partial observable Markov decision process and formulate a scheduling solution based on particle filtering. In [132], track scheduling is carried out using a modified Quality-of-Service Resource Allocation Model. Track scheduling methods have also been proposed to minimize sensor loading [133, 134]. By contrast, this work considers the scheduling of both tracking and surveillance tasks for networked radars, and quantifies both tracking and surveillance performance. In addition, the techniques presented here adaptively schedule tasks based on the characteristics of the targets within the coverage areas of the radars.

A network of phased array radars which are connected by a communication channel is considered [135]. The purpose of this work is to determine how the sharing of tracking and detection data among radars in the network can be used to enhance RRM performance. For the remainder of this chapter, the term "resource management" will refer to RRM, as opposed to the C2 concept of sensor resource management. The networked concepts developed will be referred to as Coordinated RRM, since the data from other radars is exploited in carrying out RRM. High-level concepts for Coordinated RRM will be formulated. In addition, results from the simulation of a two-radar network will illustrate the performance gains that are possible with Coordinated RRM.

5.2 PRELIMINARIES

Figure 5.1 illustrates the role of a resource manager for a single radar. This concept will be extended to networked radars. In particular, Coordinated RRM architectures and techniques will be formulated, where detection and tracking data from other radars is used in radar scheduling. In order to develop these architectures and techniques, this section presents radar network terminology and distributed tracking concepts.

5.2.1 Radar Networks

This chapter considers the resource management of a network of N monostatic radars. Although attention is restricted to monostatic radars, it is

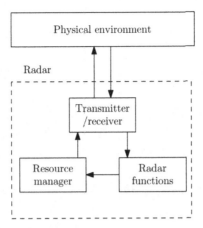

Figure 5.1 Resource management for a single radar.

possible to extend the resource management problem to include multistatic radars. Due to the large number of possible combinations of transmitters and receivers, a multistatic network will be even more complex than a monostatic network.

Different types of resource management architectures for radar networks can be formulated, and each may lead to different solutions for the resource management problem. This work considers two types of resource management architecture: centralized management and distributed management. These concepts will be specified later in this chapter. In both cases, the portion of the network that is colocated with a radar antenna will be referred to as a node.

An element common to the radar networks is a communication channel. The channel capacity, or maximum throughput, is a key element of networked radar. If the channel is wireless, the capacity will likely change over time. Resource management algorithms must therefore be able to cope with the potential of time-varying channel capacity. This work considers the case of error-free communication, as well as the case of time-varying channel availability due to errors on the communication channel.

The relationship between the coverage areas of the radar nodes is an important characteristic of the network. Consider the case when two or more nodes have coverage areas that overlap. Define the nodes with overlapping coverage areas as contributing nodes. The common coverage area will be called the overlapping region, as shown for the two-node case in Figure 5.2.

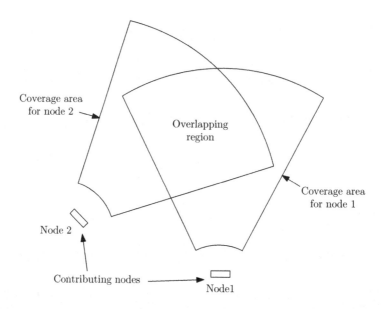

Figure 5.2 Two nodes with overlapping coverage areas.

For a tracked target or surveillance region that is located in the overlapping region, the resource manager must decide which contributing node should carry out the associated surveillance or tracking task. This assignment may vary with task and as a function of time. This adds complexity to the scheduling task for the resource manager.

If the coverage areas of each node do not overlap, then each node would be managed as in the single-radar case. If coverage areas are adjacent to each other, then tracks could be handed off from one radar to a radar with an adjacent coverage area.

5.2.2 Distributed Tracking

The extension of RRM to networked radars will build on previous results from distributed tracking in distributed sensor networks. Data association, which is the association of measurements from one or more sensors to the same target, is a key problem in multiple target tracking. When multiple sensors are connected by a communication channel, the information to be communicated on the channel must be determined. For the case of multiple hypothesis tracking, tracking performance was analyzed when a subset of hypotheses and tracks were communicated between the sensors [136]. When JPDA is used in a distributed sensor network, [137] showed that a global tracking estimate is formed by communicating the local estimates of each target along with the feasible events and their probabilities. Increasing the effective tracking update rate with a large network of track-while-scan radars was considered in [138]. A technique was presented for increasing the effective update rate while maintaining a reasonable communications bandwidth.

In general, three types of distributed tracking can be considered [139], as follows:

1. independent tracking;
2. distributed track fusion (track-to-track data association); and
3. distributed track maintenance (measurement-to-track data association).

In independent tracking, each radar conducts tracking independently of the other radars in the network, and the tracks are initiated and maintained separately. If a target is in the coverage area of multiple radars, it is likely that each radar will create a track of that target. With distributed track fusion, each radar conducts tracking independently, and track fusion is carried out via track-to-track data association to remove redundant tracks. With distributed track maintenance, a single track is created for each target, and measurement-to-track data association is conducted for measurements from all radars in the network.

5.3 ARCHITECTURE CONCEPTS FOR COORDINATED RADAR RESOURCE MANAGEMENT

Coordinated RRM includes the scheduling of tracking and surveillance tasks, the processing of tracking and detection data from other radars, and the specification of techniques for distributed tracking. As such, it addresses a time-varying multidimensional optimization problem. Since Coordinated RRM is a new area of study, this section formulates two distinct management architectures: centralized management and distributed management. Specific Coordinated RRM techniques will be implemented and analyzed later in this chapter.

5.3.1 Centralized Management Architecture

In a network with centralized management, a single resource manager formulates the schedule for all radar nodes. The resource management platform may be colocated with any one of the radar nodes or may be located

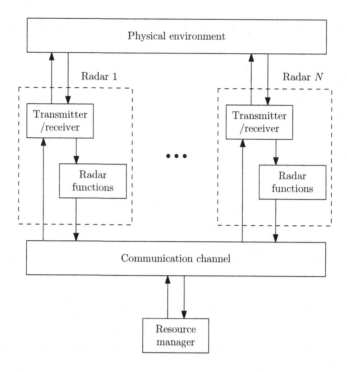

Figure 5.3 Radar network with centralized management architecture.

separately. This architecture is shown in Figure 5.3, which illustrates the case when the resource management platform is located separately from the radar nodes. The resource manager receives, on the communication channel, tracking and detection data from each of the nodes, and sends over the communication channel a resource schedule to each of the radar transmitters. The advantage of centralized management is the ability of the resource manager to utilize and control all of the available radar resources. A resource manager in a network with centralized management fully exploits the multiple radar nodes that are available. The disadvantages of centralized management include the vulnerability of varying throughput on the communication channel and the potential for data latency. If communication to any of the nodes is not available, then that radar cannot be adaptively scheduled. Each radar node could have a default resource allocation scheme that may be used in the event of failure of the communication channel.

For networks with centralized management, the resource manager must decide how to schedule N antennas at any given time. The relationship between the coverage areas of the radar nodes will affect how the antennas are scheduled. Because there is full communication between the resource manager and all of the nodes, the coverage areas of the nodes will be known at all times by the resource manager.

When overlapping regions exist and communication with one of the contributing nodes is not available, the centralized resource manager can assign one of the other contributing nodes to carry out surveillance and tracking in the overlapping region. This redundancy protects against communication failure.

5.3.2 Distributed Management Architecture

In a network with distributed management, each node is a radar that operates autonomously and has a dedicated resource manager, as shown in Figure 5.4. The resource managers communicate with each other through the communication channel. The information transmitted on the communication channel will vary depending on the resource management method that is employed. An advantage of distributed management is the reduced reliance on the communication channel, as compared to centralized management. Although the nodes are linked by the communication channel, each node is autonomous and can operate independently in the absence of communication from all other nodes. The disadvantage of distributed management is

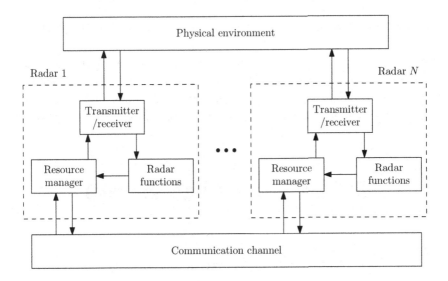

Figure 5.4 Radar network with distributed management architecture.

the distributed nature of the scheduling optimization. It is generally easier to optimize the performance of a network with centralized management.

A degenerate case of distributed management is the case where no communication channel exists. This case will be called Independent RRM and serves as a baseline against which Coordinated RRM techniques will be compared.

For networks with distributed management, each node communicates its coverage area to the other nodes in the network. If none of the nodes overlap, then each node operates independently. If nodes have adjacent coverage areas, then it may be possible to hand off tracks between the nodes.

Consider the case where overlapping regions exist. The surveillance and tracking tasks can be partitioned into overlapping tasks and exclusive tasks. Overlapping tasks are those where the associated target or surveillance region is located in an overlapping region. All other tasks are then exclusive tasks. When overlapping regions exist, a contributing node can coordinate its schedule with other contributing nodes.

For overlapping tasks, all nodes have the current estimate and relevant track information for a tracking task, and the time of the last update and detection rates for a surveillance task. The position and orientation information from other nodes allows a local node to map the received tracking and surveillance data into the local coordinate frame.

When overlapping regions exist, various types of distributed management for the contributing nodes can be specified. These are detailed in this section and are summarized in Table 5.1. The type of distributed management employed by a radar node can change with time, depending on factors including the number of contributing nodes, the size of the overlapping region, the number of overlapping tasks, and the channel capacity.

Table 5.1 Types of Distributed Management

Name	Description
Type 0	Independent management
Type 1	Autonomous management with assignment of overlapping tasks
Type 2	Autonomous management with assignment of overlapping looks
Type 3	Temporary centralized management

In Type 0 distributed management, each node carries out independent management. There is no communication between nodes. This type of distributed management will be necessary when the channel capacity is zero. Furthermore, Type 0 may be desirable when the size of the overlapping region is small relative to the individual coverage areas of the contributing nodes or when the number of overlapping tasks is small.

In Type 1 distributed management, each overlapping task is assigned to a contributing node, and all looks for that task are carried out by the assigned node. In this case, a task assignment algorithm must be developed to assign tasks to nodes. For an individual node, assigned tasks would include exclusive tasks and overlapping tasks that are assigned by the task assignment algorithm. Task assignments could change if a task ceases to be an overlapping task. It is expected that task assignment would depend on task priorities and the relative loading of the contributing nodes.

In Type 2 distributed management, individual looks corresponding to overlapping tasks are assigned dynamically to contributing nodes, so that different looks from a given task may be carried out by different nodes. For a Type 2 distributed manager, a technique for look assignment needs to be specified. Compared to Type 1 distributed management, Type 2 distributed management has increased flexibility, at the expense of increased complexity and computational costs. Look assignments will likely vary with task priorities and the relative loading of the contributing nodes. Because looks are assigned individually, it may be possible to obtain better overall performance by taking advantage of time-varying task priorities and look geometries.

In Type 3 distributed management, one of the contributing nodes is selected as the centralized manager. The manager then formulates the radar schedule for all contributing nodes. As discussed in Section 5.3.1, centralized management fully controls the resources of all contributing nodes but is vulnerable to data latency and fluctuating channel capacity. In Section 5.3.1, centralized management was specified as a system architecture where a single resource manager is used to control the radar network at all times. In Type 3 distributed management, the network has a distributed management architecture but temporarily allows one resource manager to carry out centralized management.

For looks associated with overlapping tasks, consider a technique for dynamically assigning individual looks to a contributing node. Type 2 distributed management and Type 3 distributed management provide distinct implementations of the technique, with a corresponding trade-off in computational complexity and required channel throughput. For Type 2 management, look assignments are computed at all contributing nodes. The surveillance and tracking data that is distributed among all contributing nodes is sufficient to allow all nodes to perform the same calculations, which determine look assignments. For Type 3 management, the looks assignments are computed at the centralized manager. In this case, surveillance and tracking data is distributed among all contributing nodes, and the centralized manager must also send look assignments across the channel. Compared to Type 2 management, Type 3 management requires less overall computational complexity, but increased overall channel throughput.

5.3.3 Target Prioritization for Radar Networks

Target prioritization techniques allow a radar resource manager to prioritize multiple tasks in order to develop a more effective radar schedule. In developing techniques for Coordinated RRM, the concept of target prioritization will need to be generalized to networked radars. Here, some considerations for prioritization are presented.

Fuzzy logic prioritization [13] considers a number of variables in computing a priority value for tracking tasks and surveillance tasks. For tracked targets, five variables are considered: track quality, hostility, degree of threat, weapon system capabilities, and relative position of the target.

For a given target and in the absence of communication between the nodes, the priority computed by each radar will be likely to vary. For example, the relative position of the target to each radar will probably be different. Further, if the radars are significantly separated in space, the heading and range rate, which help determine the degree of hostility, will be different for each radar. This case results in a target having a different priority relative to each radar.

An alternative approach is to compute an absolute priority for each target. The input variables for fuzzy logic prioritization can then be defined in a way that is uniform across the network. For example, the relative position could be computed relative to the radar that is closest to the target. In this case, either all radars could compute the priority using knowledge of the other radars in the network, or one radar could compute the priority and communicate the result to the other radars.

For the prioritization of surveillance sectors, four variables are considered: new targets rate (over time), number of threatening targets, threatening targets rate (over time), and original priority. For sectors that fall within the coverage area of multiple radars, it may be that the detection rate differs for each radar, due to differing clutter or noise levels, differing relative target velocities, or unfavorable aspect angles with respect to radar cross-section (RCS).

5.4 DISTRIBUTED TECHNIQUES FOR COORDINATED RADAR RESOURCE MANAGEMENT

Coordinated RRM includes the scheduling of tracking and surveillance tasks, the processing of tracking and detection data from other radars, and the specification of techniques for distributed tracking. As such, it addresses a time-varying multidimensional optimization problem. Specific scheduling techniques for a two-radar network are formulated below. For these techniques, RRM is coordinated for tracking tasks only. Surveillance tasks are conducted independently for the two radars. Errors on the communication channel may cause the channel to be unavailable for certain durations of time. This will be modeled in Section 5.4.4. For Coordinated RRM techniques, the data to be communicated between the radars will be specified.

5.4.1 Independent RRM

In this case, each radar carries out Independent RRM for all tasks. This was referred to as Type 0 management in Table 5.1 and is the baseline case against which Coordinated RRM will be assessed. No data is communicated between the radars. Each radar utilizes an independent tracker and employs Independent RRM that includes three aspects of adaptivity:

1. fuzzy logic prioritization;
2. adaptive track update intervals; and
3. time-balancing scheduling.

The fuzzy logic prioritization technique [13] is implemented for tracking tasks. For each tracked target, characteristics such as heading, range, range rate, height, and maneuver history are used to compute a target priority value between 0 and 1. In this way, the relative priority of each tracked target is assessed, so that more radar resources can be assigned to higher-priority targets.

The tracker requests an update interval for each tracked target, and this request is sent to the scheduler. The requested track update interval depends on the target priority as follows:

$$\text{Requested track update interval} = \begin{cases} 1.5 \text{ s}, & \text{if target priority} \geq 0.75 \\ 3 \text{ s}, & \text{if target priority} < 0.75 \end{cases},$$

(5.1)

where the target priority is a value between 0 and 1. If the track updates are scheduled at their requested intervals, then targets with a priority greater than 0.75 are updated twice as frequently as lower-priority targets.

The scheduling of tracking and surveillance tasks is conducted using the time-balancing scheduler [40, 127]. Each task has an associated time balance. If a look associated with that task is not scheduled, then the task time balance increases linearly with time. If a look is scheduled, the time balance decreases. At any given time, the task with the highest time balance is scheduled next.

5.4.2 Type 1 Management

When the channel is available, Type 1 management assigns overlapping tracking tasks to the radar that has the smaller range to the tracked target. Once the overlapping task has been assigned to a radar, that radar carries out all track updates until the track ends. An overview of the assignment rules for tracking tasks is shown in Figure 5.5. Each radar conducts surveillance over its entire coverage area. Each radar also conducts tracking of its exclusive tracking tasks.

For assigned tracking tasks, the fuzzy logic algorithm is used to compute the relative priorities of each tracked target. Adaptive track update intervals are computed using (5.1). Surveillance looks and tracking looks are then scheduled using the time-balancing scheduler.

Detection-to-track association is carried out for all tracks, including tracks assigned to the other radar. For example, assume that track y is assigned to Radar 1. In the course of conducting surveillance, a detection by Radar 2 will be gated against all tracks, include that of track y. If the detection is gated to track y, then the detection will be used to update track y. If the detection is not gated to track y, then Radar 1 schedules a track confirmation look.

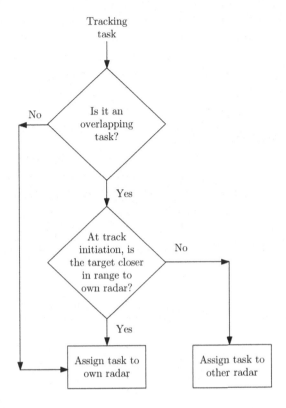

Figure 5.5 Task assignment algorithm for Type 1 management.

For Type 1 management, the data sent across the communication channel is specified in Table 5.2. The position, velocity, and orientation of each radar platform are sent to the other platform, so that both radars can compute coverage areas and the overlapping region, if any. This data also allows detections from the other radar to be mapped into the local coordinate frame. The estimated position of targets at track confirmation is required to compute the task assignment algorithm. Once an overlapping tracking task has been assigned to a particular radar, only detections in the overlapping region are sent across the channel.

Table 5.2 Data Sent Across the Communication Channel for Type 1 Management	
Platform	**Overlapping Tasks**
Position	Detections
Velocity	Estimated position at track confirmation
Orientation	

In Type 1 management, overlapping tasks are not assigned to both radars, which reduces the time required for tracking tasks compared to Independent RRM. In particular, the radar that is not assigned to a particular track does not assign looks to update that track, which frees up the radar to carry out other tasks. The benefit gained from the coordinated scheduling of overlapping tasks will be quantified in Section 5.5.

5.4.3 Type 2 Management

When the channel is available, Type 2 management assigns overlapping tracking tasks to a radar on a look-by-look basis. Each look is assigned to the radar that has the smaller range to the tracked target. An overview of the assignment rules for tracking looks is shown in Figure 5.6. Note that Type 2 management is computationally more intensive than Type 1 management, because a comparison of the target ranges to each radar is carried out

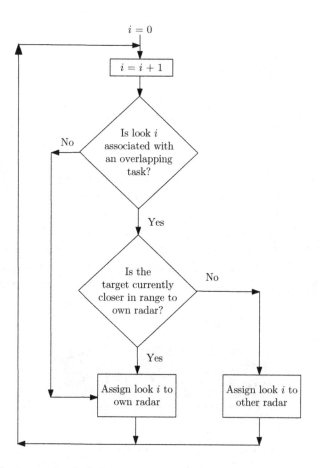

Figure 5.6 Look assignment algorithm for Type 2 management, for looks $i = 1, 2, \ldots$ of a given tracking task.

for each look associated with a tracking task. Each radar carries out surveillance of its entire coverage area and conducts tracking of its exclusive tracking tasks.

After each tracking look has been scheduled, the next look is assigned to a radar based on minimum range. The fuzzy logic priority (relative to the assigned radar) and the adaptive track update interval are computed. Surveillance looks and assigned tracking looks are scheduled for each radar using the time-balancing scheduler. As was the case with Type 1 management, detection-to-track association is carried out for all tracks, including tracks assigned to the other radar.

For Type 2 management, the data sent across the communication channel is specified in Table 5.3. The position, velocity, and orientation of each radar platform are sent to the other platform, so that both radars can compute coverage areas and the overlapping region, if any. Detections and tracks associated with overlapping tasks are required, since the estimated range to each radar is used to compute the look assignment on a look-by-look basis. A given track may be updated by either radar, using scheduled track update looks or detections from surveillance looks that are gated with the track.

5.4.4 Model for Communication Channel Availability

To implement Coordinated RRM techniques, the radar network relies on a communication channel between radars to transmit and receive data related to target detections and tracks. It is assumed that the radar network employs a digital communication system with Forward Error Correction (FEC) channel coding [140]. If the Bit Error Rate (BER) of the channel is less than or equal to the maximum BER of the FEC code, then the data is received without error. However, if the BER of the channel is greater than the maximum BER of the FEC code, then the data is not received reliably.

This work models the effects of errors on the communication channel, together with error control coding employed by the communication system. When the BER of the channel is less than or equal to the maximum BER of

Table 5.3 Data Sent Across the Communication Channel for Type 2 Management	
Platform	**Overlapping Tasks**
Position	Detections
Velocity	Tracks
Orientation	

the FEC code, then the channel is available. When the BER of the channel is greater than the maximum BER of the FEC code, then the channel is not available. Over time, the channel is available with probability p. This realistic model for channel availability accounts for errors that may occur due to interference on the channel, together with error control coding that would be employed by the communication system.

5.5 TWO-RADAR NETWORK EXAMPLE

Section 5.4 formulated techniques for Coordinated RRM. In this section, a two-radar network example is considered, and the performance of these techniques is analyzed. The performance analysis utilizes the Adapt_MFR simulation tool, which was described in Chapter 3.

The scenario is shown in Figure 5.7 and is specified as follows. The two radars are stationary and are separated by 10 km, with the second radar located directly south of the first radar. The boresites of both radars point directly east. Each radar is capable of scanning ±60° in azimuth.

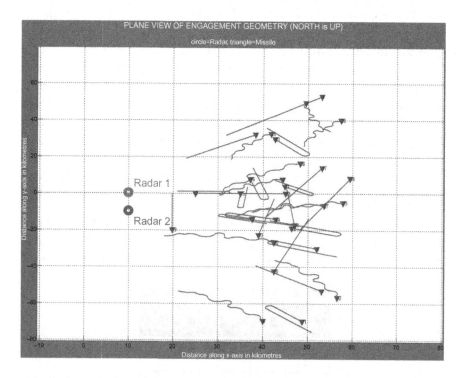

Figure 5.7 Top-down view of radar positions and target trajectories for the scenario with Target Set A. Triangles indicate target position at the start of its trajectory.

Table 5.4 Set of Parameter Values for 30 Targets		
Parameter	Values: Target Set A	Values: Target Set B
Altitude (m)	500, 600, 750	500, 600, 750
Velocity (m/s)	100, 150	200, 250
Radar cross-section (m²)	50, 75	5, 10
Trajectory	Straight line, U-turn, Weave	Straight line, U-turn, Weave

The scenario consists of 30 targets with trajectories defined over a time interval of 200 s. Each target has a fixed altitude, RCS, and velocity. In addition, each target follows one of three trajectory types. The targets have varying values of initial position and initial heading, which are chosen so that each target trajectory is within the azimuthal coverage extent of one or both radars for the entire time interval.

Two sets of targets are considered: Target Set A and Target Set B. The parameter values for the target sets are listed in Table 5.4. It is seen that Target Set B has targets with smaller RCS and larger velocity values. Figure 5.7 shows a top-down view of the radar locations and target trajectories for Target Set A.

Adapt_MFR simulations were run for the scenario with Target Set A. The following five cases were considered, where p is the probability of channel availability, as described in Section 5.4.4.

1. Independent RRM;
2. Type 1 management with $p = 1$;
3. Type 2 management with $p = 1$;
4. Type 1 management with $p = 0.5$; and
5. Type 2 management with $p = 0.5$.

An IMM tracker with NN-JPDA [111] was utilized in all cases. The track initiation process is as follows. After a target detection, the radar specifies a target confirmation look for that target. If the target is confirmed, then a tentative track is formed. After a tentative track has been updated two times in three attempts, the tentative track becomes a confirmed track. For the purposes of computing track occupancy, track confirmation looks are associated with target detection, while update looks for tentative tracks or confirmed tracks are associated with target tracking.

For the case of Type 1 management with $p = 1$, Figure 5.8 shows the number of tracks with priority greater than or equal to 0.75, and the number of tracks with priority less than 0.75. Both are plotted against simulation time for each radar. The priority of a track determines the requested track update interval, as specified in (5.1). The total number of tracks may not

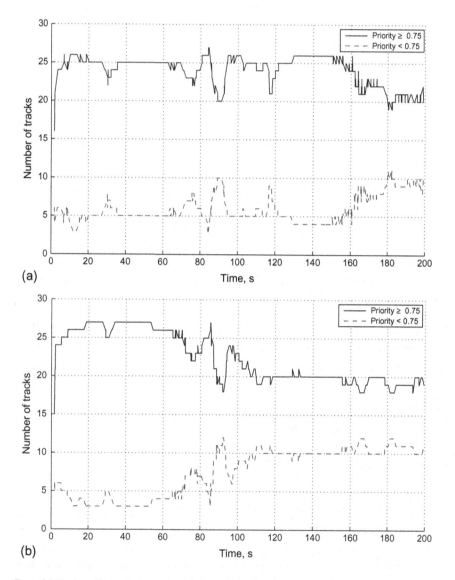

Figure 5.8 Number of high-priority and low-priority tracks for Type 1 management for Target Set A. (a) Radar 1. (b) Radar 2.

always equal the number of targets, 30, because at certain brief periods of time during the simulation, there may be untracked targets or false tracks.

For $p = 1$, the communication channel was available during the entire simulation. For $p = 0.5$, the simulation time interval of 200 s was divided into subintervals of 10 s. For each subinterval, the channel was randomly chosen as either being available or not available, with equal probability. For Type 1 management with $p = 0.5$, a transition from the channel being available to not available resulted in the two radars initiating new tracks independently. When the channel transitioned from being not available to available, multiple tracks of the same target were fused into a single track. For Type 2 management with $p = 0.5$, a transition from the channel being available to not available required that existing tracks be assigned to one of the radars. Each track was assigned to the radar that most recently updated the track. As was the case with Type 1 management, when the channel transitioned from not available to available, multiple tracks of the same target were fused into a single track. Track-to-track association was carried out using target ground truth to associate multiple tracks with each target. Track-to-track fusion was then performed using an averaging scheme, which resulted in only one track being associated with each target. In a real-world environment, track-to-track association and fusion could be carried out statistically [139, pp. 195-197].

Figure 5.9 shows track completeness for the six cases of Independent RRM—Radar 1, Independent RRM—Radar 2, Type 1 management with $p = 1$, Type 2 management with $p = 1$, Type 1 management with $p = 0.5$, and Type 2 management with $p = 0.5$. Track completeness was computed as specified in Chapter 3. For Independent RRM, tracking is carried out independently for the two radars. The results for Type 1 consider any track that is associated with a given target, regardless of which radar was assigned the track. The results for Type 2 includes tracked targets where updates were carried out by a single radar and those where updates were carried out by both radars, as per the look assignment specified in Figure 5.6. The results indicate that targets are tracked with track completeness of 0.95 or greater, with the exception of Target 4, whose trajectory is shown in Figure 5.10. Target 4 starts at a longer range and travels toward Radars 1 and 2. With Independent RRM, Target 4 is not tracked by Radar 2 until later in the scenario, due to lower signal-to-noise ratio at the start of the scenario. This accounts for the track completeness of 0.82 for Independent RRM—Radar 2.

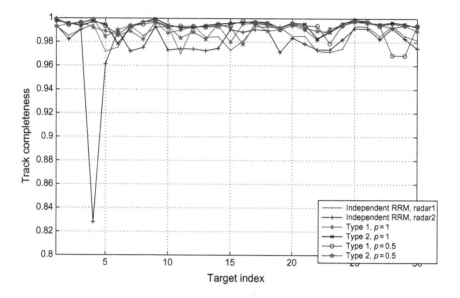

Figure 5.9 *Track completeness for the scenario with Target Set A.*

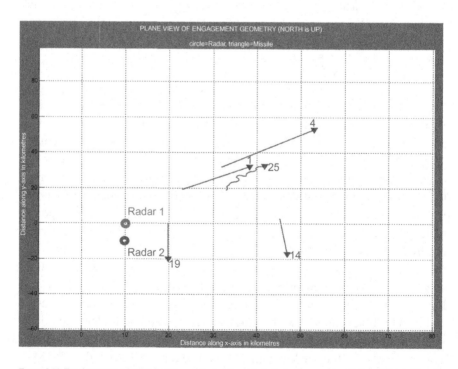

Figure 5.10 *Top-down view of radar locations and select target trajectories for scenario with Target Set A. Triangles indicate target position at the start of its trajectory.*

Track occupancy results for both radars are presented in Figure 5.11. For Type 1 management and Type 2 management, tracks associated with targets in the overlapping region are updated by only one of the two radars when the communication channel is available. For Independent RRM, such tracks are updated by both radars, which increases track occupancy for both radars. For fixed $p = 1$ or $p = 0.5$, Type 1 management and Type 2 management

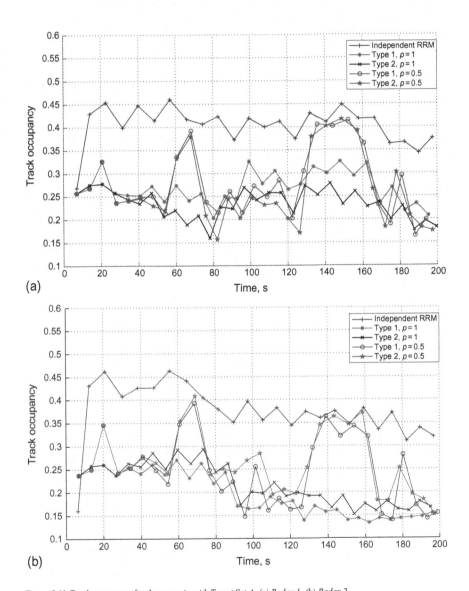

(a)

(b)

Figure 5.11 Track occupancy for the scenario with Target Set A. (a) Radar 1. (b) Radar 2.

have similar track occupancy values. Type 1 management carries out task assignment for overlapping tasks, while Type 2 management carries out look assignment for overlapping tasks. The distinction between task assignment and look assignment has a negligible effect on track occupancy. The tooth-like structure of the track occupancy plots is caused by slight variations in the number of track updates in consecutive fixed intervals. During intervals when the channel is not available, the track occupancy of Type 1 with $p = 0.5$ and Type 2 with $p = 0.5$ increase to that of the Independent RRM case, as expected. This can be seen during the intervals from 50 to 70 s and from 130 to 160 s.

The decreased track occupancy resulting from the use of Coordinated RRM increases the time available for surveillance. This results in decreased frame time for both radars, as shown in Figure 5.12. Compared to Independent RRM, the frame time for Type 1 management, $p = 1$ and Type 2 management, $p = 1$ is decreased by approximately 2 s. As a result, the reaction time against new threats is improved. As expected, the frame time for Type 1 management, $p = 0.5$ and Type 2 management, $p = 0.5$ increases to that of Independent RRM when the channel is not available. These results apply to the 30-target scenario under consideration. For a scenario with a larger number of targets in the overlapping region, the frame time for all cases would increase. However, the difference in frame time between Independent RRM and Coordinated RRM would also increase, indicating a more significant advantage for Coordinated RRM.

Figure 5.13 plots the difference in position error between Type 2 management with $p = 1$ and Type 1 management with $p = 1$, for all 30 targets in Target Set A. Positive difference corresponds with lower Type 2 error. For some targets, Type 1 management has a smaller position error, while Type 2 management has smaller position error for other targets. For this target scenario, neither the use of Type 1 or Type 2 management results in smaller estimation error. For a small number of targets, there are periods of time when the estimation error has sharp increases in value for either Type 1 or Type 2 management, which causes a spike in the difference value plotted in Figure 5.13. The increase in estimation error value occurs when two or more targets cross paths, and the tracker momentarily associates the track with a different target.

Figure 5.14 compares track occupancy for Target Set A with that for Target Set B. Figure 5.14a and b shows track occupancy for Type 1 management with $p = 1$ for Radars 1 and 2. Although the track occupancy

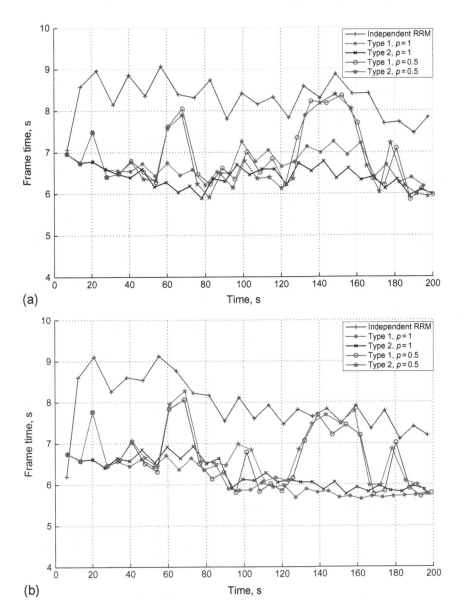

Figure 5.12 Frame time for the scenario with Target Set A. (a) Radar 1. (b) Radar 2.

is similar for Radar 2, Target Set B has somewhat lower track occupancy for Radar 1. This is because the targets in Target Set B are moving away from the radars at a higher velocity, which decreases target priority and increases track update intervals. For Type 2 management with $p = 1$, Figure 5.14c and d shows track occupancy for Radars 1 and 2. Again in this case, track

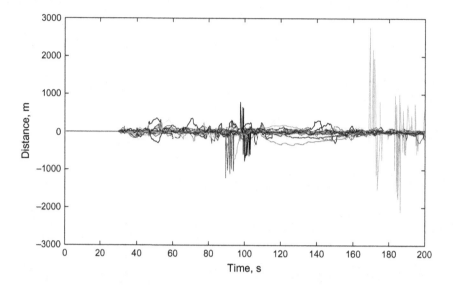

Figure 5.13 Difference between position error for Type 2 management with p = 1 and position error for Type 1 management with p = 1. Positive difference corresponds with lower Type 2 error.

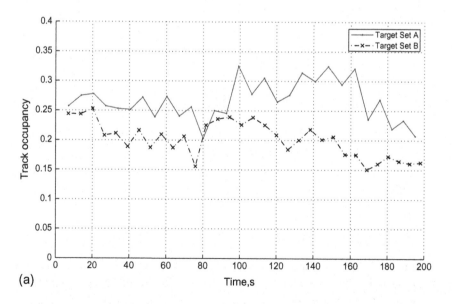

Figure 5.14 Comparison of track occupancy for Target Set A and Target Set B. (a) Radar 1: Type 1 management, p = 1.

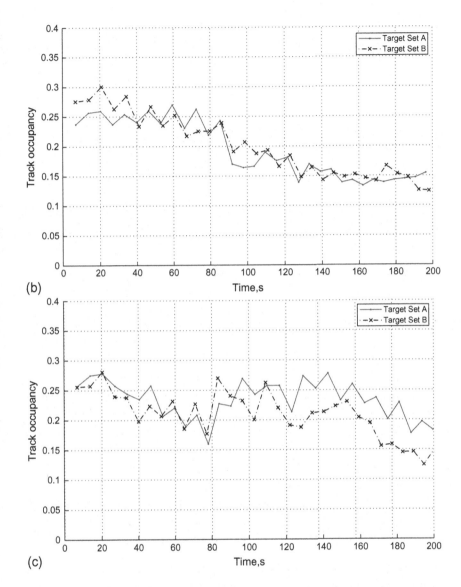

Figure 5.14, Cont'd (b) Radar 2: Type 1 management, p = 1. (c) Radar 1: Type 2 management, p = 1.

occupancy is similar for Radar 2, but Target Set B has slightly lower track occupancy for Radar 1. Similar to Type 1, this is caused by higher-velocity targets that are moving away from the radars.

Results from the 30-target scenario show that Type 1 management and Type 2 management achieve track completeness close to one, with similar results for Independent RRM. However, when the communication channel

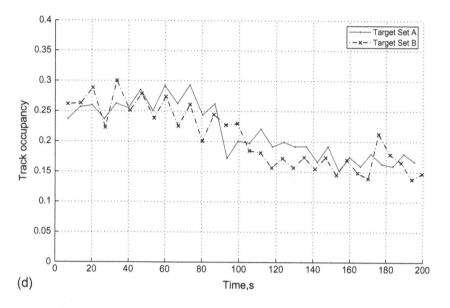

Figure 5.14, Cont'd (d) Radar 2: Type 2 management, p = 1.

is available, Type 1 management and Type 2 management have decreased track occupancy and decreased frame time compared to Independent RRM. This indicates that a radar network using Coordinated RRM can improve reaction time against new threats. To achieve this enhanced tracking performance, the radars must send data across the communication channel. The data to be transmitted includes the position, velocity, and orientation of each radar platform, detections associated with overlapping tasks, and the estimated position of targets at track confirmation. In addition, for Type 2 management, tracks associated with overlapping tasks must be transmitted. When the communication channel is not available, results showed that the performance of Coordinated RRM is similar to that of Independent RRM.

A radar is overloaded when not all tracking look requests can be scheduled. In this case, it is likely that track completeness will not be one for all targets. Coordinated RRM can improve track completeness compared to Independent RRM when the individual radars are overloaded. Overall, differences in track completeness and track occupancy between Type 1 management and Type 2 management will depend on the task assignment and look assignment algorithms.

5.6 SUMMARY

This chapter considered whether the sharing of detection and tracking data can enhance RRM performance. Coordinated RRM exploits data that is transmitted across a communication channel. Two types of Coordinated RRM techniques were formulated, with each type characterized by varying amounts of coordination between the radar nodes. A 2-radar network and 30-target scenario were modeled in the simulation tool Adapt_MFR, to analyze the performance of Independent RRM and Coordinated RRM. All RRM techniques utilized adaptive task prioritization, track update intervals, and radar scheduling. It was shown that Coordinated RRM achieves the same track completeness as Independent RRM, while decreasing track occupancy and frame time. Therefore, Coordinated RRM can improve reaction time against threats, at the expense of sending data across a communication channel. The performance of Coordinated RRM for a communication channel with errors was also modeled and analyzed. For the examples considered here, there was no difference in performance between Type 1 management and Type 2 management.

The use of Coordinated RRM offers the potential for significant performance improvements; however, the analysis of further radar and target scenarios is required before definitive conclusions can be drawn about the benefits of Coordinated RRM and about comparisons between Type 1 management and Type 2 management. The example in Section 5.5 utilized RRM techniques based on fuzzy logic prioritization and the time-balancing scheduler. Independent RRM and Coordinated RRM based on other techniques, such as those presented in Chapter 2, should also be considered.

CHAPTER *6*

Conclusions

This chapter summarizes the key results from Chapters 1 to 5. In Section 6.1, some common themes of the book are presented. Section 6.2 describes future work in radar resource management (RRM).

6.1 COMMON THEMES

There are a number of common themes that appear throughout this book. First, it is clear that RRM is challenging because of the time-varying nature of radar scheduling. Many of the algorithms that were described in Chapter 2 attempt to formulate optimal scheduling solutions, but are extremely complex because the optimization is carried out for the entire timeline of radar operation. Due to this high computational complexity, optimal techniques such as dynamic programming and neural network algorithms cannot be implemented in real-time schedulers. The challenge of implementing a real-time scheduler has led to the development of numerous suboptimal techniques, such as the Optimal Assignment Scheduler and the Sequential Scheduler. However, even these techniques are somewhat complex, and the TSBF Scheduler also requires tuning of the benefit function parameters.

Second, RRM techniques that are adaptive can modify their behavior in response to changing environments. Chapter 3 considered Adaptive RRM employing fuzzy logic prioritization, time-balancing scheduling, and adaptive track update rates. Performance was compared to that of Nonadaptive RRM, and it was shown that Adaptive RRM has similar track completeness, but with lower track occupancy and lower frame time. Adaptive RRM is able to allocate resources to tracking while using less radar timeline. The Coordinated RRM techniques presented in Chapter 5 also illustrate the advantages of adaptivity. Compared to independent RRM, coordinated RRM had the same track completeness with lower track occupancy and

Adaptive Radar Resource Management. http://dx.doi.org/10.1016/B978-0-12-802902-2.00006-5

137

lower frame time. This was due to the ability of coordinated RRM to allocate overlapping tasks adaptively to a single radar.

Finally, it is clear that optimizing tracking performance is a key element of RRM. The objective of the radar is to detect and track all targets within its field of regard. A number of performance metrics were introduced in Chapter 3. Track completeness, track accuracy, and tracking delays were analyzed extensively throughout the book, and the performance of various adaptive RRM techniques was quantified through the analysis of these metrics. The importance of measuring tracking performance motivated the development of Adapt_MFR and the implementation of the IMM tracker. This simulation approach allows a number of target scenarios to be modeled in Adapt_MFR. The operation of the IMM tracker then allows for the generation of tracks, which can be compared to simulation ground truth.

6.2 FUTURE WORK

Based on the survey of the open literature, the following topics are suggested:

- study of adaptive classification algorithm for RRM;
- comparison of the fuzzy logic, neural network, and entropy algorithms;
- application of fuzzy logic for task scheduling;
- evaluation of the dynamic programming and Q-RAM algorithms with realistic RRM problems;
- investigation of the waveform diversity benefits for RRM; and
- study of the motion noise models for adaptive update rate tracking.

The Benchmark 3 problem should also be studied and future solutions should be tested and compared against the existing solution. Additional sensors should be considered to enhance the RRM performance. Other measures of performance (MOP) such as task occupancy and timeliness should be included.

Although this book has described a number of new results in RRM, there are still many challenges that remain in this important area of study. First, there is a need to establish a set of benchmark problems for RRM. As described in Chapter 2, NRL introduced benchmark problems for tracking that led to a number of advances in tracking. Benchmark problems for RRM would specify common radar parameters and target scenarios that enable researchers to evaluate their algorithms against a common framework. These benchmark problems would necessarily be very complex, to capture the time-varying nature of the RRM problem and to specify an environment that forces an RRM algorithm to adjust its behavior in the presence of a changing environment.

Second, more RRM techniques should be evaluated against realistic target scenarios. In Chapter 3, Adaptive RRM included fuzzy logic prioritization, time-balancing scheduling, and adaptive track update rates. The performance of Adaptive RRM was then compared to Nonadaptive RRM using 52-target and 152-target scenarios. As described in Chapter 2, there are numerous techniques for task prioritization and task scheduling. Many of these techniques could be implemented in Adapt_MFR and analyzed against multiple-target scenarios. Analysis of the tracking performance associated with various RRM techniques provides insight into the characteristics of those techniques.

Finally, further study is required on the interaction between radar scheduling and waveform selection. Waveform selection has been considered by waveform-aided algorithms, which were described in Section 2.4. However, joint adaptive scheduling and adaptive waveform selection has not been previously considered. Such techniques may have high complexity with the potential for enhanced detection and tracking performance.

BIBLIOGRAPHY

[1] T. Jeffrey, Phased Array Radar Design, SciTech, Raleigh, NC, 2009.

[2] S. Sabatini, M. Tarantino, Multifunction Array Radar: System Design and Analysis, Artech House, Boston, 1994.

[3] E. Brookner, Phased arrays and radar: past, present and future, Microw. J. 49 (1) (2006) 24-46.

[4] A.G. Huizing, A.A.F. Bloemen, An efficient scheduling algorithm for a multifunction radar, in: IEEE International Symposium on Phased Array Systems and Technology, 1996, pp. 359-364, doi:10.1109/PAST.1996.566115.

[5] R. Rajkumar, C. Lee, J. Lehoczky, D. Siewiorek, A resource allocation model for QoS management, in: Proceedings of the IEEE Real Time Systems Symposium, 1997, pp. 290-307.

[6] C. Lee, J. Lechoczky, D. Siewiorek, On QoS optimization with discrete QoS options, in: Proceedings of the IEEE Real Time and Embedded Technology and Applications Symposium, 1998, pp. 276-286.

[7] W. Komorniczak, J. Pietrasinski, Selected problems of MFR resources management, in: Proceedings of the 3rd International Conference on Information Fusion, Paris, France, 2000, pp. 3-8.

[8] W. Komorniczak, T. Kuczerski, J. Pietrasinski, The priority assignment for detected targets in multifunction radar, Res. J. Telecommun. Inf. Technol. 1 (2001) 30-32.

[9] A. Izquierdo-Fuente, J.R. Casar-Corredera, Optimal radar pulse scheduling using a neural network, in: Proceedings of the IEEE World Congress on Neural Networks and Computational Intelligence, 1994, pp. 4588-4591.

[10] V. Vannicole, Expert system for sensor resource allocation, in: Proceedings of the IEEE International Radar Conference, 1991, pp. 1005-1008.

[11] J.F. Pietrasinski, W. Komorniczak, Application of artificial intelligence in radar resource management, in: Proceedings of the 12th International Conference on Microwaves and Radar, 1990, pp. 138-142.

[12] M.T. Vine, Fuzzy logic in radar resource management, in: IEE Multifunction Radar and Sonar Sensor Management Techniques, 2001, pp. 1-4.

[13] S.L.C. Miranda, K. Baker, K. Woodbridge, H.D. Griffiths, Fuzzy logic approach for prioritisation of radar tasks and sectors of surveillance in multifunction radar, IET Proc. Radar Sonar Navig., vol. 1, 2007, pp. 131-141.

[14] A.P. Stoffel, Heuristic energy management for active array multifunction radars, in: Proceedings of the IEEE National Telesystems Conference, San Diego, 1994, pp. 71-74.

[15] W. Komorniczak, J. Pietrasinski, B. Solaiman, The data fusion approach to the priority assignment in the multifunction radar, in: Proceedings of the 14th Conference on Microwave, Radar and Wireless Communication, 2002, pp. 647-650.

[16] S.L.C. Miranda, K. Baker, K. Woodbridge, H.D. Griffiths, Knowledge-based resource management for multifunction radar, IEEE Signal Process. Mag. 66 (1) (2006) 66-76.

[17] S.L.C. Miranda, K. Baker, K. Woodbridge, H.D. Griffiths, Simulation methods for prioritizing tasks and sectors of surveillance in phased array radar, J. Simul. 5 (1-2) (2005) 18-25.

[18] S.L.C. Miranda, K. Baker, K. Woodbridge, H.D. Griffiths, Phased array radar resource manage-
    ment: a comparison of scheduling algorithms, in: Proceedings of the IEEE International Radar
    Conference, 2004, pp. 79-84.

[19] K. Woodbridge, C. Baker, Tracking optimization for multifunction radar, in: Proceedings of the
    London Communications Symposium, 2002.

[20] B. Dawber, Fuzzy logic module, TTCP PowerPoint Presentation, 2005.

[21] P.E. Berry, D.A.B. Fogg, On the use of entropy for optimal radar resource management and
    control, in: Proceedings of the IEEE International Radar Conference, 2003, pp. 572-577.

[22] A.G. Huizing, J.A. Spruyt, Adaptive waveform selection with a neural network, in: Proceedings
    of the IEEE International Radar Conference, 1992.

[23] B.L. Scala, B. Moran, Optimal target tracking with restless bandits, Digital Signal Process. 16
    (2005) 479-487.

[24] V. Krishnamurthy, R.J. Evans, Hidden Markov model multiarm bandits: a methodology for beam
    scheduling in multitarget tracking, IEEE Trans. Signal Process. 49 (2) (2001) 2893-2908.

[25] J. Wintenby, V. Krishnamurthy, Hierarchical resource management in adaptive airborne surveil-
    lance radars, IEEE Trans. Aerosp. Electron. Syst. 42 (2) (2006) 401-420.

[26] J. Wintenby, Resource allocation in airborne surveillance radar, Ph.D. dissertation, Chalmers
    University of Technology, Sweden, 2003.

[27] D. Stromberg, P. Grahn, Scheduling of tasks in phased array radar, in: IEEE International
    Symposium on Phased Array Systems and Technology, 1996.

[28] M. Elshafei, H.D. Sherali, J.C. Smith, Radar pulse interleaving for multi-target tracking, Nav.
    Res. Logist. 5 (1) (2003) 72-94.

[29] R.B. Washburn, M.K. Schneider, J.J. Fox, Stochastic dynamic programming based approaches
    to sensor resource management, in: Proceedings of the International Conference on Information
    Fusion, Annapolis, MD, 2002, pp. 608-615.

[30] S. Howard, S. Suvorova, B. Moran, Optimal policy for scheduling of Gauss-Markov systems, in:
    Proceedings of the International Conference on Information Fusion, Stockholm, Sweden, 2004,
    pp. 888-892.

[31] A.J. Orman, C.N. Potts, A.K. Shahani, A.R. Moore, Scheduling for a multifunction phased array
    radar system, Eur. J. Oper. Res. 90 (1) (1996) 13-25.

[32] A.J. Orman, A.K. Shahani, A.R. Moore, Modelling for the control of complex radar system,
    Comput. Oper. Res. 25 (3) (1998) 239-249.

[33] A.J. Orman, C.N. Potts, On the complexity of coupled-task scheduling, Discret. Appl. Math. 72
    (1) (1997) 141-154.

[34] J.M. Butler, A.R. Moore, H.D. Griffiths, Resource management for a rotating MFR, in: Proceed-
    ings of the IEEE International Radar Conference, 1997, pp. 568-572.

[35] C. Duron, J.M. Proth, Multifunction radar: task scheduling, J. Math. Model. Algorithms (1) (2002)
    105-116.

[36] C. Duron, J.M. Proth, Linked task scheduling: algorithms for the single machine case, Tech. Rep.,
    2002.

[37] C. Duron, J.M. Proth, Insertion of a random bitask in a schedule: a real-time approach, Comput.
    Oper. Res. 31 (2004) 779-790.

[38] E. Winter, L. Lupinski, On scheduling the dwells of a multifunction radar, in: Proceedings of the
    IEEE International Conference on Radar, 2006, pp. 1-4.

[39] R. Filippi, S. Pardini, An example of resources management in a multifunctional rotating phased array radar, in: Proceedings of the IEE Colloquium on Real-Time Management of Adaptive Radar Systems, 1990, pp. 1-3.

[40] J.M. Butler, Multi-function radar tracking and control, Ph.D. thesis, University College London, 1998.

[41] C.F. Kuo, T.W. Kuo, C. Chang, Real-time digital signal processing of phased array radars, IEEE Trans. Parallel Distrib. Syst. 145 (2003) 433-446.

[42] C.G. Lee, P.S. Kang, C.S. Shih, L. Sha, Schedulability envelope for real-time radar dwell scheduling, IEEE Trans. Comput. 55 (12) (2006) 1599-1613.

[43] T.W. Kuo, A.S. Chao, C.F. Kuo, C. Chang, Y. Su, Real-time digital signal processing of phased array radars, in: Proceedings of the IEEE International Radar Conference, 2002, pp. 160-171.

[44] C. Shih, S. Gopalakrishnan, P. Ganti, M. Caccamo, L. Sha, Scheduling real-time dwells using tasks with synthetic periods, in: Proceedings of the IEEE Real Time Systems Symposium, 2003, pp. 210-219.

[45] C. Shih, S. Gopalakrishnan, P. Ganti, M. Caccamo, L. Sha, Template-based real time dwells scheduling with energy constraints, in: Proceedings of the IEEE Real Time and Embedded Technology and Applications Symposium, 2003, pp. 19-27.

[46] S. Goddard, K. Jeffray, Analyzing the real-time properties of a dataflow execution paradigm using a synthetic aperture radar application, in: Proceedings of the IEEE Real Time and Embedded Technology and Applications Symposium, 1997, pp. 60-71.

[47] S. Ghosh, Scalable QoS-based resource allocation, Ph.D. thesis, Carnegie Mellon University, 2004.

[48] S. Ghosh, R. Rajkumar, J. Hansen, J. Lehoczky, Integrated QoS-aware resource management and scheduling with multi-resource constraints, Real-Time Syst. 33 (2006) 7-46.

[49] J.P. Hansen, S. Ghosh, R. Rajkumar, J. Lehoczky, Resource management of highly configurable tasks, in: Proceedings of the 18th International Parallel and Distributed Processing Symposium, 2004, pp. 140-147.

[50] S. Gopalakrishnan, C.S. Shih, P. Ganti, M. Caccamo, L. Sha, Radar dwell scheduling with temporal distance and energy constraints, in: Proceedings of the IEEE International Radar Conference, 2004, pp. 1-34.

[51] S. Gopalakrishnan, M. Caccamo, C.S. Shih, C. Lee, L. Sha, Finite-horizon scheduling of radar dwells with online template construction, Real-Time Syst. 33 (1) (2006) 47-75.

[52] K. Harada, T. Ushio, Y. Nakamoto, Adaptive resource allocation control for fair QoS management, IEEE Trans. Comput. 56 (3) (2007) 344-357.

[53] D.J. Kershaw, R.J. Evans, Waveform selected PDA, IEEE Trans. Aerosp. Electron. Syst. 33 (4) (1997) 1180-1188.

[54] D.J. Kershaw, R.J. Evans, Optimal waveform selection for tracking systems, IEEE Trans. Inf. Theory 40 (5) (1994) 492-496.

[55] S. Howard, S. Suvorova, B. Moran, Optimal policy for scheduling of Gauss-Markov systems, in: Proceedings of the 7th International Conference on Information Fusion, Stockholm, Sweden, 2004, pp. 888-892.

[56] S. Suvorova, S.D. Howard, W. Moran, Beam and waveform scheduling approach to combined radar surveillance and tracking the paranoid tracker, in: Proceedings of the International Waveform Diversity and Design Conference, Hawaii, USA, 2006.

[57] B.L. Scala, B. Moran, R. Evands, Optimal adaptive waveform selection for target detection, in: Proceedings of the IEEE International Radar Conference, Adelaide, Australia, 2003, pp. 492-496.

[58] B.L. Scala, M. Rezaeian, B. Moran, Optimal adaptive waveform selection for target tracking, in: Proceedings the 8th International Conference on Information Fusion, Philadelphia, USA, 2005, pp. 25-28.

[59] S.M. Sowelam, A.H. Tewfik, Waveform selection in radar target classification, IEEE Trans. Inf. Theory 46 (3) (2000) 1014-1029.

[60] S. Howard, S. Suvorova, B. Moran, Waveform libraries for radar tracking applications, in: Proceedings of the 1st International Conference on Waveform Diversity and Design, Edinburgh, UK, 2004, pp. 1424-1428.

[61] B. Moran, S. Suvorova, S. Howard, Sensor management for radar: a tutorial, in: Proceedings of the Sensing for Security Workshop, Ciocco, Italy, 2005, pp. 1-23.

[62] F. Harada, T. Ushio, Y. Nakamoto, Adaptive resource allocation control for fair QoS management, IEEE Trans. Comput. 56 (3) (2007) 344-357.

[63] S. Suvorova, D. Musicki, B. Moran, S. Howard, B.L. Scala, Multi step ahead beam and waveform scheduling for tracking of maneuvering targets in clutter, in: Proceedings of International Conference Acoustics, Speech, Signal Processing (ICASSP), Philadelphia, USA, 2005, pp. 889-892.

[64] A. Leshem, O. Naparstek, A. Nehorai, Information theoretic adaptive radar waveform design for multiple extended targets, IEEE J. Sel. Top. Sign. Process. 1 (1) (2007) 42-55.

[65] S. Sira, A. Papandreou-Suppappola, D. Morrell, Dynamic configuration of time-varying waveforms for agile sensing and tracking in clutter, IEEE Trans. Signal Process. 55 (7) (2007) 3207-3217.

[66] S. Sira, D. Cochran, A. Papandreou-Suppappola, D. Morrell, W. Moran, S. Howard, R. Calderbank, Adaptive waveform design for improved detection of low-RCS targets in heavy sea clutter, IEEE J. Sel. Top. Sign. Process. 1 (1) (2007) 56-66.

[67] M. Hurtado, T. Zhao, A. Nehorai, Adaptive polarized waveform design for target tracking based on sequential Bayesian inference, IEEE Trans. Signal Process. 56 (3) (2008) 1120-1133.

[68] S. Haykin, B. Currie, T. Kirubarajan, Literature search on adaptive radar transmit waveforms, Tech. Rep., 2003.

[69] DARP Adaptive Waveform Design, http://signal.ese.wustl.edu/DARPA/publications.html, Accessed Nov. 25, 2008.

[70] F. Daum, R. Fitzgerald, Decoupled Kalman filters for phased array radar tracking, IEEE Trans. Autom. Control 28 (3) (1983) 296-283.

[71] G. Keuk, S. Blackman, On phased array radar tracking and parameter control, IEEE Trans. Aerosp. Electron. Syst. 29 (1) (1993) 186-194.

[72] W. Koch, On adaptive parameter control for phased-array tracking, in: Proceedings of the SPIE Conference on Signal and Data Processing of Small Targets, vol. 3809, 1999, pp. 444-455.

[73] H.J. Shin, Adaptive-update-rate target tracking for phased-array radar, IEE Radar Sonar Navig. 142 (3) (1995) 137-143.

[74] H. Leung, A Hopfield neural tracker for phased array antenna, IEEE Trans. Aerosp. Electron. Syst. 33 (1) (1997) 301-307.

[75] H. Sun-Mog, J. Young-Hun, Optimal scheduling of track updates in phased array radars, IEEE Trans. Aerosp. Electron. Syst. 34 (3) (1998) 1016-1022.

[76] K. Tei-Wei, C. Yung-Sheng, C.-F. Kuo, C. Chang, Real-time dwell scheduling of component-oriented phased array radars, IEEE Trans. Comput. 54 (1) (2005) 47-60.

[77] G. Keuk, Software structure and sampling strategy for automatic target tracking with phased array radar, in: Proceedings of the AGARD, 1978, pp. 21-32.

[78] S.A. Cohen, Adaptive variable update rate algorithm for tracking targets with a phased array radar, IEE Proc. Radar Sonar Navig. 133 (3) (1986) 277-280.

[79] V.C. Vannicola, J.A. Mineo, Expert system for sensor resource allocation, in: The 33rd Midwest Symposium on Circuits and Systems, 1990, pp. 1005-1008.

[80] M. Mune, M. Harrison, D. Wilkin, M.S. Woolfson, Comparison of adaptive target-tracking algorithm for phased array radar, IEE Radar Sonar Navig. 139 (5) (1992) 336-342.

[81] P.W. Sarunic, Adaptive variable update rate target tracking for a phased array radar, in: Proceedings of the IEEE International Radar Conference, 1995, pp. 317-322.

[82] P.W. Sarunic, R.J. Evans, Adaptive update rate tracking using IMM nearest neighbour algorithm incorporating rapid re-looks, IEE Radar Sonar Navig. 144 (4) (1997) 195-204.

[83] S.P. Noyes, Calculation of next time for track update in the MESAR phased array radar, in: IEEE Colloquium on Target Tracking and Data Fusion, 1998, pp. 2/1-2/7.

[84] T.W. Jeffrey, Phased array radar tracking with non-uniformly spaced measurements, in: Proceedings of the IEEE International Radar Conference, 1998, pp. 44-49.

[85] S.L. Coetzee, K. Woodbridge, C.J. Baker, Multifunction radar resource management using tracking optimization, in: Proceedings of the IEEE International Radar Conference, 2003, pp. 578-583.

[86] S. Gopalakrishnan, C.S. Shih, P. Ganti, M. Caccamo, L. Sha, Radar dwell scheduling with temporal distance and energy constraints, in: Proceedings of the IEEE International Radar Conference, 2004.

[87] C.G. Lee, A novel framework for quality-aware resource management in phased array radar systems, in: Proceedings of the 11th IEEE Real Time and Embedded Technology and Applications Symposium, 2005, pp. 322-331.

[88] J.H. Zwaga, Y. Boers, H. Driessen, On tracking performance constrained MFR parameter control, in: Proceedings of the 6th International Conference on Information Fusion, Cairns, Queensland, Australia, 2003, pp. 712-718.

[89] J.H. Zwaga, H. Driessen, Tracking performance constrained MFR parameter control: applying constraints on prediction accuracy, in: Proceedings of the 8th International Conference on Information Fusion, Philadelphia, USA, 2005, pp. 25-28.

[90] Y. Boers, H. Driessen, J. Zwaga, Adaptive MFR parameter control: fixed against variable P_d, IEE Proc. Radar Sonar Navig. 153 (1) (2006) 2-6.

[91] C.G. Lee, C.S. Shih, L. Sha, Schedulability envelope for real-time radar dwell scheduling, IEEE Trans. Comput. 55 (12) (2006) 1599-1613.

[92] H. Benoudnine, M. Keche, A. Ouamri, M.S. Woolfson, Fast adaptive update rate for phased array radar using IMM target tracking algorithm, in: Proceedings of the IEEE International Symposium on Signal Processing and Information Technology, 2006, pp. 277-282.

[93] W.D. Blair, G.A. Watson, S.A. Hoffman, Benchmark problem for beam pointing control of phased array radar against maneuvering targets, in: Proceedings of the American Control Conference, Baltimore, USA, 1994, pp. 2071-2075.

[94] W.D. Blair, G.A. Watson, T. Kirubarajan, Y. Bar-Shalom, Benchmark for radar resource allocation and tracking in the presence of ECM, IEEE Trans. Aerosp. Electron. Syst. 34 (4) (1998) 1097-1114.

[95] W.D. Blair, G.A. Watson, IMM algorithm for solution to benchmark problem for tracking maneuvering targets, in: Proceedings of the SPIE Acquisition, Tracking and Pointing Conference, Orlando, USA, 1994, pp. 476-488.

[96] G.A. Watson, W.D. Blair, Tracking performance of a phased array radar with revisit time controlled using the IMM algorithm, in: Proceedings of the American Control Conference, Baltimore, USA, 1994, pp. 160-165.

[97] E. Daeipour, Y. Bar-Shalom, L. Li, Adaptive beam pointing control of a phased array radar using an IMM estimator, in: Proceedings of the American Control Conference, Baltimore, USA, 1994, pp. 2093-2097.

[98] T. Kirubarajan, Y. Bar-Shalom, E. Daeipour, Adaptive beam pointing control of phased array radar using an IMM estimator, in: Proceedings of the American Control Conference, Baltimore, USA, 1994, pp. 2616-2620.

[99] W.D. Blair, G.A. Watson, S.A. Hoffman, Benchmark problem for beam pointing control of phased array radar against maneuvering targets in the presence of ECM and false alarms, in: Proceedings of the American Control Conference, Seattle, USA, 1995, pp. 2601-2605.

[100] S.S. Blackman, M. Bush, G. Demos, R. Popoli, IMM/MHT tracking and data association for benchmark tracking problem, in: Proceedings of the American Control Conference, Seattle, USA, 1995, pp. 2606-2610.

[101] P. Kalata, An alpha-beta target tracking approach to the benchmark tracking problem, in: Proceedings of the American Control Conference, Baltimore, USA, 1994, pp. 2076-2080.

[102] T. Kirubarajan, Y. Bar-Shalom, E. Daeipour, Adaptive beam pointing control of a phased array radar in the presence of ECM and false alarms using IMMPDAF, in: Proceedings of the American Control Conference, Seattle, USA, 1995, pp. 2616-2620.

[103] S.A. Hoffman, W.D. Blair, Guidance, tracking and radar resource management for self defense, in: Proceedings of the IEEE Conference on Decision and Control, New Orleans, USA, 1995, pp. 2772-2777.

[104] M. Efe, D.P. Atherton, Adaptive beam pointing control of a phased array radar using the AIMM algorithm, in: Proceedings of the IEE Colloquium on Target Tracking and Data Fusion, 1996, pp. 11/1-11/8.

[105] R.E. Popoli, S.S. Blackman, M.T. Busch, Application of multiple hypothesis tracking to agile beam radar tracking, in: Proceedings of the SPIE Conference on Signal and Data Processing of Small Targets, 1996, pp. 418-428.

[106] T. Kirubarajan, Y. Bar-Shalom, W.D. Blair, G.A. Watson, IMMPDAF for radar management and tracking benchmark with ECM, IEEE Trans. Aerosp. Electron. Syst. 34 (4) (1998) 1115-1134.

[107] D. Angelova, E. Semerdjiev, L. Mihaylova, X. Li, An IMMPDAF solution to benchmark problem for tracking in clutter and standoff jammer, Inf. Secur. 2 (1999) 1-8.

[108] P.D. Burns, W.D. Blair, Optimal phased array radar beam pointing for MTT, in: Proceedings of Aerospace Conference, 2004, pp. 1851-1858.

[109] G.A. Watson, D.H. McCabe, Benchmark problem with a multisensor framework for radar resource allocation and tracking of highly maneuvering targets, closely-spaced targets, and targets in the presence of sea-surface-induced multipath, Tech. Rep., 1999, Technical Report NSWC-DD/TR-99/32, NSWC, Dahlgren, VA.

[110] A. Sinha, T. Kirubarajan, Y. Bar-Shalom, Tracker and signal processing for the benchmark problem with unresolved targets, IEEE Trans. Aerosp. Electron. Syst. 42 (1) (2006) 279-300.

[111] R. Helmick, IMM estimator with nearest-neighbor joint probabilistic data association, chap. 3, in: Multitarget-Multisensor Tracking: Applications and Advances, Artech House, 2000 .

[112] C. Duron, J.M. Proth, Insertion of a random bitask in a schedule: a real-time approach, Comput. Oper. Res. 31 (2004) 779-790.

[113] A.J. Orman, C.N. Potts, A.K. Shahani, A.R. Moore, Scheduling for a multifunction phased array radar system, Eur. J. Oper. Res. 90 (1) (1996) 13-25.

[114] E. Winter, L. Lupinski, On scheduling the dwells of a multifunction radar, in: International Conference on Radar, Shanghai, China, 2006.

[115] J.M. Butler, A.R. Moore, H.D. Griffiths, Resource management for a rotating MFR, in: Proceedings of the IEEE International Radar Conference, 1997, pp. 568-572.

[116] W.K. Stafford, Real time control of multifunction electronically scanned adaptive radar, in: IEE Colloquium on Real Time Management of Adaptive Radar Systems, 1990.

[117] G.T. Capraro, A. Farina, H. Griffiths, M.C. Wicks, Knowledge-based radar signal and data processing, IEEE Sig. Proc. Mag. (2006) 18Ü29.

[118] D.P. Bertsekas, An auction algorithm for shortest paths, SIAM J. Optim. 1 (1991) 425-447.

[119] Y. Bar-Shalom, X.R. Li, T. Kirubarajan, Estimation with applications to tracking and navigation, John Wiley and Sons, New York, 2001.

[120] G.B. Dantzig, Recent advances in linear programming, Manag. Sci. 2 (1956) 131-144.

[121] R. Fourer, A simplex algorithm for piecewise-linear programming I: derivation and proof, Math. Program. 33 (1985) 204-233.

[122] F. Güder, F.J. Nourie, A dual simplex algorithm for piecewise-linear programming, J. Oper. Res. Soc. 47 (1996) 583-590.

[123] R. Fourer, A simplex algorithm for piecewise-linear programming III: computational analysis and applications, Math. Program. 53 (1992) 213-235.

[124] W. Press, S.A. Teukolski, W. Vetterling, B.P. Flannery, Numerical Recipes in C, second ed., Cambridge University Press, Cambridge, 1992.

[125] S.P. Noyes, Calculation of next time for track update in the MESAR phased array radar, in: IEE Colloquium on Target Tracking and Data Fusion, Digest No. 1998/282, 1998, pp. 1-7.

[126] G. Davidson, Cooperation between tracking and radar resource management, in: IET International Conference on Radar Systems, 2007, pp. 1-4.

[127] S.L.C. Miranda, C.J. Baker, K. Woodbridge, H.D. Griffiths, Comparison of scheduling algorithms for multifunction radar, IET Radar Sonar Navig. 1 (6) (2007) 414-424.

[128] D.A. Pierre, Optimization theory with applications, Dover, New York, 1986.

[129] B.W. Johnson, J.M. Green, Naval network-centric sensor resource management, in: 7th International Command and Control Research and Technology Symposium (ICCRTS), www.dtic.mil, 2002.

[130] Y. He, E.K.P. Chong, Sensor scheduling for target tracking in sensor networks, in: 43rd IEEE Conference on Decision and Control (CDC), vol. 1, ISSN 0191-2216, 2004, pp. 743-748, doi: 10.1109/CDC.2004.1428743.

[131] Y. He, E.K.P. Chong, Sensor scheduling for target tracking: a Monte Carlo sampling approach, Digital Signal Process. 16 (2006) 533-545.

[132] A. Charlish, Tasking networked multi-function radar systems for active tracking, in: 14th International Radar Symposium (IRS), June, vol. 1, 2013, pp. 367-374.

[133] A.S. Narykov, A. Yarovoy, Sensor selection algorithm for optimal management of the tracking capability in multisensor radar system, in: 2013 European Radar Conference (EuRAD), 2013, pp. 499-502.

[134] A.S. Narykov, O.A. Krasnov, A. Yarovoy, Algorithm for resource management of multiple phased array radars for target tracking, in: 16th International Conference on Information Fusion (FUSION), 2013, pp. 1258-1264.

[135] Y. Teng, H.D. Griffiths, C.J. Baker, K. Woodbridge, Netted radar sensitivity and ambiguity, IET Radar Sonar Navig. 1 (6) (2007) 479-486, ISSN 1751-8784, doi:10.1049/iet-rsn:20070005.

[136] C.-Y. Chong, K.-C. Chang, S. Mori, Distributed Tracking in Distributed Sensor Networks, in: American Control Conference, 1986, pp. 1863-1868.

[137] K.-C. Chang, C.-Y. Chong, Y. Bar-Shalom, Joint probabilistic data association in distributed sensor networks, IEEE Trans. Autom. Control 31 (10) (1986) 889-897, ISSN 0018-9286, doi: 10.1109/TAC.1986.1104143.

[138] G.W. Deley, A netting approach to automatic radar track initiation, association, and tracking in air surveillance systems, in: G. Vankeuk (Ed.), AGARD Strategies for Autom. Track Initiation 10 p (SEE N79-30454 21-32), 1979.

[139] Y. Bar-Shalom, Multitarget-multisensor tracking: advanced applications, Artec House, Norwood, MA, 1990.

[140] S. Lin, D.J. Costello, Error Control Coding: Fundamentals and Applications, Prentice-Hall, Englewood Cliffs, NJ, 1983.

Printed in the United States
By Bookmasters